SPACE RACE

The U.S.-U.S.S.R. Competition to Reach the Moon

Martin J. Collins and the Division of Space History

National Air and Space Museum ◆ The Smithsonian Institution

Pomegranate

SAN FRANCISCO

Published by Pomegranate Communications, Inc.
Box 6099, Rohnert Park, CA 94917
(800) 227-1428; (707) 586-5500; www.pomegranate.com

Pomegranate Europe Ltd.
Fullbridge House, Fullbridge
Maldon, Essex CM9 4LE, England

Pomegranate Catalog No. A940
ISBN 0-7649-0905-3

Library of Congress Cataloging-in-Publication Data

Collins, Martin, 1951–
    Space race : the U.S.–U.S.S.R. competition to reach the moon / Martin J. Collins and the Division of Space History, National Air and Space Museum, the Smithsonian Institution.
        p.   cm.
    ISBN 0-7649-0905-3 (pbk.)
    1. Space race—United States—Exhibitions. 2. Space race—Soviet Union—Exhibitions. 3. Space flight to the moon—Exhibitions.
    I. National Air and Space Museum. Division of Space History.
    II. Title.
    TL788.5.C65 1999
    629.45'4'09—dc21   98-47057
                                                                                         CIP

Cover and interior design by Monroe Street Studios

Printed in Korea

08 07 06 05 04 03 02 01 00 99   10 9 8 7 6 5 4 3 2 1

Contents

Acknowledgments

Space Race was a richly cooperative venture. Within the Museum, the exhibit benefited from the work of many. The Division of Space History collaborated in the writing of the script, much of which is presented in this catalog. Exhibit designer William Jacobs transformed the Division's ideas, words, and artifacts into a practical and elegant presentation. Smithsonian photographer Eric Long expertly captured the finished exhibit through numerous shots of the gallery and artifacts. Many of his photographs are in this book. James David, Division of Space History, gave crucial assistance in gathering copyright permissions.

The efforts of Museum staff in preparing Space Race would not have been possible without the support of the Perot Foundation and Ross Perot. Through auction, they acquired many of the Soviet space artifacts that enrich the exhibit, then arranged for their loan to the Museum. When paired with their American counterparts, these artifacts are at the heart of the story. Emmet Stephenson and Art Dula also loaned important Soviet artifacts. The Museum of the Yuri Gagarin Cosmonauts Training Center, Star City, Russia, graciously loaned several artifacts relating to the career of Cosmonaut Gagarin. The Perot Foundation also provided support in exhibit preparations and working with Soviet officials. The Ruth and Julius Wile Foundation gave a major donation without which the exhibit could not have gone forward. The Department of Defense Legacy Program supported research during exhibit preparation. The Rocket Space Corporation Energia of Russia offered access to historical resources and invaluable collaboration.

Introduction

"What makes the Soviet threat unique in history is its all-inclusiveness. Every human activity is pressed into service as a weapon of expansion. Trade, economic development, military power, arts, science, education, the whole world of ideas. . . . The Soviets are, in short, waging total cold war."

—President Dwight D. Eisenhower, 1958

"Finally, if we are to win the battle that is now going on around the world between freedom and tyranny, the dramatic achievements in space which occurred in recent weeks should have made clear to us all, as did the Sputnik in 1957, the impact of this adventure on the minds of men everywhere."

—President John F. Kennedy, 1961

"We have a long way to go in the space race. We started late. But this is the new ocean, and I believe the United States must sail on it and be in a position second to none."

—President John F. Kennedy, 1962

On 4 October 1957, a Soviet R-7 intercontinental ballistic missile launched Sputnik, the first human-made object to be placed in Earth orbit. In the tense atmosphere of the Cold War, the event fueled the fears and imaginations of people around the world and sparked the most dramatic technological competition the world has ever seen.

Space Race, a new permanent exhibition at the Smithsonian's National Air and Space Museum, opened in May 1997. The gallery examines the spectacular, publicly celebrated milestones of our first steps into space, as well as highly secret efforts to spy on adversaries from high above the Earth. This book is a companion piece to the exhibition, but it also stands alone to tell the compelling story of the Cold War contest between the United States and the Soviet Union.

The Space Race exhibit was made possible by a climactic event: the end of the Cold War. The collapse of the Soviet Union and the waning of decades of antagonism have opened central aspects of space race history

to examination for the first time. American and Soviet artifacts are now available that speak to defining events in the race to the moon, in the military use of space, and in the lives of the men and women who carried the aspirations of each country. Through these artifacts, the exhibit outlines the background and motivation that shaped both the American and the Soviet venture into space.

The Cold War left an indelible imprint on the United States and the Soviet Union. More than forty years of confrontation defined life in each place in profound ways. Before Sputnik in 1957, an arms race, threat of nuclear annihiliation, espionage and counter espionage, a war in Korea, and a continuous clashing of words and ideas at home and around the world had already made the Cold War part of the fabric of individual lives and of the two nations. After Sputnik, the intensity of the Cold War conflict continued, reflected in events such as the building of the Berlin Wall, the Cuban missile crisis, and the Vietnam War. The first steps into space and the

Left: The Space Race gallery, National Air and Space Museum

moon race that followed—representing some of the most dramatic achievements of the century—were inseparable from this context. Space became another arena of competition, shaped by the larger course of the Cold War and altering its direction.

The history of U.S. efforts in space has several intertwining threads. Astronauts' and cosmonauts' highly public feats and the race to the moon are the best-known aspects of early space exploration. They provided an image of space as a story of explorers and the conquest of new frontiers. This image built on long-standing conceptions of exploration as a motif in American history. Many advocates of the moon race promoted U.S. efforts as a natural extension of our early history of exploration.

The remarkable firsts in space achieved by astronauts were the fruits of a civilian space program headed by the National Aeronautics and Space Administration (NASA). President Dwight Eisenhower chartered NASA in Sputnik's wake to promote the peaceful and scientific exploration of space. NASA's Mercury, Gemini, and Apollo flights came to be identified by the press and the public as our space program. But soon after Sputnik, Eisenhower created two national security space programs. One was a military space program, primarily implemented by the Air Force to exploit space as a vantage from which to defend against and to wage war. The other was a top-secret effort to gather intelligence from orbiting satellites on the closed society of the Soviet Union and its allies. This was undertaken by the Central Intelligence Agency, the Air Force, and a new organization called the National Reconnaissance Office, whose existence was classified as top secret until recent years. These space programs coexisted throughout the space race; they still persist as separate efforts. Each represented a different response to the challenges of the Cold War.

We now know that this reconnaissance effort was Eisenhower's first priority in space. With the exception of the Apollo era, presidents and congresses have appropriated larger budgets for military and intelligence programs than for those of NASA. Interest in spying, weather observation, and science from space

dates to the immediate post-World War II period, when the military services, particularly the Air Force, began to explore the feasibility of rockets and satellites. Sputnik only helped to catalyze and redirect Cold War efforts already underway.

It now seems commonplace for space to be a frontier of exploration and a crucial vantage for national security, but it took extraordinary circumstances to start these programs. Soon after World War II, our jarring Cold War confrontation with the Soviet Union provided the basic ingredient: a massive mobilization of scientific and technological resources throughout the country, in government, industry, and universities. Our political traditions set the formula: the federal government provided policy and organization, while private enterprise provided the means. Through government contracts, the productive engines of private industry and the expertise of universities built the necessary technologies. Space exploration was not just a triumph of individuals but of bureaucracies, institutions, and a political system marshaled to meet the Soviet challenge. Exploration by astronauts, cosmonauts, and automated satellites was the visible tip of a vast institutional and political response. Human explorers stood as symbols and proxies for their societies. Military and reconnaissance programs represented the high stakes of exploiting space for national security.

This complex reaction to Sputnik reflected the world's perception that the first satellite was at once a political, technological, and military act. The first human-made object in the heavens was not for exploration; it was a clear indication that the Soviets possessed strong, reliable rockets that could reach American soil with nuclear warheads. Gathering intelligence on Soviet military activities suddenly seemed more urgent.

The Soviet Union saw Sputnik as a statement that its technology, and by implication its political system, were superior and more worthy of emulation by developing countries. Soviets and Americans perceived space exploration, particularly human spectaculars, as a crucial symbolic arena for demonstrating the superiority of their respective political systems. With

These soaring, if somewhat abstract, political calculations soon yielded to the clamor of the race. On the American side, every phase of the enterprise spawned excitement and new experiences—and riveted the attention of the press, especially the burgeoning medium of television. Astronauts, in particular, embodying the "right stuff," pulled their fellow citizens along on a vicarious journey into space.

People could imagine the dangers of spaceflight, of piloting a spacecraft back into the atmosphere from space.

President Kennedy's 1961 call to land a man on the moon before the end of the decade, the race to the moon became the ultimate test of the two superpowers' standing in the world.

It is hard to convey the intensity of public concern generated by Sputnik. The press, Congress, and people in the street saw the Soviet challenge as an issue of great national moment. Newspaper headlines, political cartoons, and congressional hearings worried over the loss of American international prestige as the Soviets led the way into space. By the time of cosmonaut Yuri Gagarin's first human flight into space in April 1961, U.S. leaders were ready to commit to a moon journey to trump the USSR in the struggle for the hearts and minds of people around the world. In the context of the time, the reasons seemed self-evident. As expressed in a background paper for Kennedy's famous May 1961 speech, "dramatic achievements in space . . . symbolize the technological power and organizing capacity of a nation. Our attainments are a major element in the international competition between the Soviet system and our own . . . . Lunar and planetary exploration are, in this sense, part of the battle along the fluid front of the cold war."

As author Tom Wolfe put it, the astronaut's life was on the line. He had to have "the moxie, the reflexes, the experience, the coolness, to pull [his craft] back at the last yawning moment."[1] Confronting the looming unknowns of space travel, each detail of the astronaut experience was fresh and compelling—home life, training, every step of a flight. Astronaut Michael Collins related his experience sitting on top of a Saturn V rocket, the largest ever built, during the Apollo 11 launch to the moon. "We are thrown left and right against our straps in spasmodic little jerks. It is steering like crazy, like a nervous lady driving a wide car down a narrow alley"[2]—the effect of four and a half million pounds of propellant shooting out the rocket's aft in just 150 seconds. Alan Shepard, John Glenn, Neil Armstrong, Saturn V rockets, and Apollo capsules—the people and machines all seemed gigantic and heroic, especially when viewed against the backdrop of their relentless Soviet counterparts.

---

[1] Tom Wolfe, *The Right Stuff* (New York: Farrar, Straus, and Giroux, 1979).

[2] Michael Collins, *Carrying the Fire: An Astronaut's Journeys* (New York: Farrar, Straus, and Giroux, 1974)

Thousands of engineers around the country got in on the excitement, too. Building space hardware and taking it to the launch site provided another arena of experience for the public and the press to appreciate. One engineer recalled his work in launching some of first versions of the Saturn rocket. "You remained on the pad as the liquid oxygen prechilled, with xenon lights, and the wind blowing, and as those pipes chill, they scream. The vents are blowing . . . this thing is groaning and moaning and the hydraulic pumps are coming on. . . . At the moment of ignition, hearing that sharp crack . . . we would watch that thing ignite with a beautiful, absolute, thunderous roar, zillions of horsepower, and you visualize them valves working and them turbo pumps going ch-ch-ch-ch . . . . Clearing the pad . . . the thing is smoking and venting and shaking and screaming."[3]

Space Race endeavors to capture the rich mix of politics, culture, and individual experience that shaped the Soviet and American competition. The exhibit covers the three distinct threads of the venture into space—military origins, reconnaissance, and exploration—beginning with a look at the military and scientific rocket programs that started the journey into space.

After World War II, the United States and the Soviet Union engaged in a tense rivalry to design rockets powerful enough to send thermonuclear weapons across the globe. Building on technology gleaned from captured German V-2 missiles, by the 1950s both sides had succeeded in creating the weapon that epitomized Cold War fear: the intercontinental ballistic missile. Some new rockets were readily adapted to launch satellites into orbit and humans into space. What began as a competition to build bomb-carrying missiles expanded into a contest to reach space. Rockets in the Air and Space Museum collection, from a World War II German V-2 to the Minuteman III (the mainstay of the U.S. ballistic-missile arsenal), tell the history of this period before the space race.

The story of the space race is the exhibit's core, offering a unique opportunity to see side-by-side Soviet

and American artifacts that detail the spectacular feats leading to landings on the moon. The end of the Cold War has cast fresh light on Soviet activities. For much of the 1960s, the Soviets denied that they were in a race to the moon. Displayed in the United States for the first time, the Space Race exhibit contains artifacts that reveal the extent of the Soviet effort to land humans on the moon first. A Krechet space suit designed for lunar excursions is displayed next to astronaut David Scott's Apollo 15 lunar suit, the knees and shins embedded with moondust. Reproductions from the personal diaries of Vasili Mishin, a dominant figure in the Soviet space program, provide insight into Soviet planning and activities. These diaries offer compelling technical details and observations on the Soviet scene, including the failure of the N-1 rocket, the crucial factor in losing the race to the moon. Equally dramatic are artifacts associated with Soviet firsts in space—a battery key from Sputnik, Yuri Gagarin's space suit, and a space suit and airlock used in training for Aleksei Leonov's walk in space. Leonov's color-pencil drawing of the planet's horizon provides a touching indication of the emotional impact of viewing Earth from space.

The presence of Soviet artifacts in the Space Race exhibit is a sharp reminder of post-Cold War chaos and hard times in Russia and other republics. These national treasures were auctioned through Sotheby's. Collectors in the United States, most notably the Perot Foundation, acquired numerous items and made them available to the Museum on loan. The Perot Foundation aims to repatriate these artifacts to the Russian people.

On the United States side, we can now tell the story of one of the earliest, greatest, and most secret of U.S. space successes—a program code-named Corona. Under Corona, U.S. officials developed reconnaissance satellites that carried increasingly sophisticated cameras to photograph the Soviet Union from space. In 1995 the Central Intelligence Agency and the National Reconnaissance Office transferred a KH-4B camera, the most advanced in the Corona spy program, to the Museum. At the same time, thousands of images taken from 1960 to 1972 were declassified. Corona emphasized surveillance of the Soviet military,

[3] Sylvia D. Fries, *NASA Engineers in the Age of Apollo* (Washington, D.C.: NASA, 1992)

but it also provided crucial information on progress and failures in the Soviet space program. A photograph of a Soviet launch pad after an explosion of the N-1 rocket reveals the U.S. capability to track its adversary's space preparations.

Corona's most important contribution was to provide for the first time regular, detailed information on the state of the Soviet military. Before Corona, the United States had limited and sporadic means to gather intelligence on the Soviet Union. The best-known program was the U-2 spy plane, which was shot down over Soviet territory in May 1960. In August 1960, Corona made its first successful reconnaissance flight. The photographs gave U.S. leaders confidence that, despite Soviet missile and space successes, America did not lag behind. Such information eased Cold War tensions and later made possible treaties to limit increases in nuclear weapon stockpiles on each side. Reconnaissance satellites helped to assure that the "balance of terror" created by missiles and nuclear weapons did not tip into a hot war.

A lively spirit of cooperation marked the

exhibit's preparation. Curators, the Perot Foundation, and Russian officials worked together. An introductory video features astronaut Thomas Stafford and cosmonauts Gherman Titov and Aleksei Leonov discussing the years of competition and the first cooperative U.S.–U.S.S.R. venture in space, Apollo-Soyuz. The centerpiece of the exhibit's opening was a salute to pioneer astronauts and cosmonauts. Thirteen astronauts and nine cosmonauts shared the stage and their recollections. These efforts also symbolized the end of the Cold War and the space race and the beginning of an era of cooperation between the former rivals.

Through the many stories told by American and Soviet artifacts, *Space Race* reminds us that the Cold War fundamentally shaped our venture into space. The accomplishments we associate with the space program—landing on the moon, sending scientific probes to planets in our solar system, and achieving human flights into orbit via the Space Shuttle—were part of this larger frame of reference. *Space Race* tells of the time when our leaders made the heavens inseparable from the earth.

CHAPTER 1

Military Origins

"The next war will not start with a naval action nor . . . by aircraft flown by human beings. It might very well start with missiles being dropped on the capital of a country, say Washington."

—*Gen. Henry H. Arnold, 1945*

After World War II, the rocket foreshadowed a new style of warfare in which nuclear bombs could be delivered quickly across the world. War might begin—and end—suddenly, decisively, without warning.

As the space race began, the United States and the Soviet Union were building rockets to use as long-range weapons. The United States initially favored bombers, but the Soviets preferred missiles and thus took an early lead in rocket technology.

A rocket that could carry a bomb across the globe could also be used to loft machines and men into orbit. The United States and the Soviet Union engaged in a long competition to develop rockets for both warfare and space exploration.

Three types of rockets played important roles in research, national defense, or space exploration. A missile delivers an explosive warhead to a target. A sounding rocket carries scientific instruments into the upper atmosphere. A launch vehicle sends spacecraft into Earth orbit or beyond.

The First Ballistic Missile

V-2, or Vengeance Weapon 2 (Vergeltungswaffe Zwei), was the name Nazi propagandists gave to the first ballistic missile used to strike distant targets. German Army Ordnance had been developing rocketry since the 1930s, aiming to create a long-range missile and exploring the use of rocket-powered aircraft. The liquid-propellant V-2 missile was first flown successfully from Peenemünde, Germany, on the Baltic Sea in October 1942.

Late in World War II, Germany launched nearly three thousand V-2s against England, France, and Belgium. After the war, the United States and the Soviet Union used captured V-2s as a basis for developing their own large rockets.

The German army recruited a young engineer, Wernher von Braun, shown here briefing officers in 1943 at Peenemünde, where the V-2 was developed. Von Braun quickly became the technical director of Germany's long-range missile program.

On page 15 (left to right):

Aerobee 150 (slim gray-and-black rocket) was one of a family of sounding rockets used from 1947 to 1985 for upper-atmosphere research.

Vanguard (tall black-and-silver rocket) boosted the second U.S. satellite, *Vanguard 1,* into space on March 17, 1958.

Jupiter-C (large white rocket marked "UE") placed the first U.S. satellite, Explorer 1, into orbit on 31 January 1958.

Minuteman III (large green rocket)—this intercontinental ballistic missile (ICBM) has been a U.S. strategic weapon since 1970.

Viking (white rocket with silver top) was a U.S. Navy refinement of the V-2, used in the late 1940s and early 1950s as a sounding rocket and test vehicle for Project Vanguard.

Scout-D (tall white rocket marked "United States") was used by NASA and other customers from 1961 to 1994 to launch small scientific satellites.

WAC *Corporal* (small white rocket with black top)—this high-altitude U.S. Army test rocket, used from 1945 to 1950, represented the state of American rocketry at the time of the German V-2.

V-2 (black-and-white-patterned rocket) was developed by the German army during World War II. It inspired large-scale rocketry in the United States and Soviet Union, and set the stage for long-range ballistic missiles.

(Overhead) *V-1*—a cruise missile, known as a "Buzz Bomb," used by the Germans in World War II.

Wernher von Braun at Peenemünde

The V-2 was the largest and most complex missile in the German arsenal. It could send over a ton of explosives more than 150 miles downrange in five minutes.

The single rocket engine used a mixture of alcohol and liquid oxygen to provide thrust for about a minute. After engine shutoff, the missile traveled to its target on a ballistic trajectory—that is, falling under the influence of gravity.

During powered flight, the V-2 was guided either by radio signals from the ground or by onboard gyroscopes and a device that measured the rocket's acceleration. It had control vanes in the rocket exhaust and air vanes on its fins.

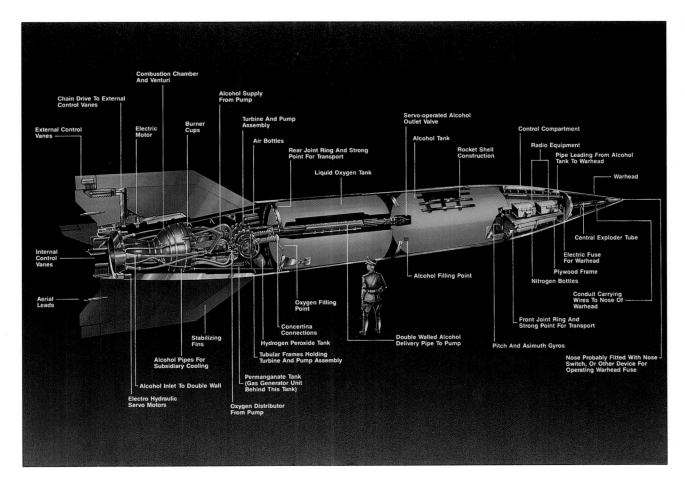

Combustion Chamber
And Venturi

Chain Drive To External
Control Vanes

Alcohol Supply
From Pump

Servo-operated Alcohol
Outlet Valve

Control Compartment

External Control
Vanes

Electric
Motor

Burner
Cups

Turbine And Pump
Assembly

Alcohol Tank

Radio Equipment

Air Bottles

Rear Joint Ring And Strong
Point For Transport

Rocket Shell
Construction

Pipe Leading From Alcohol
Tank To Warhead

Liquid Oxygen Tank

Warhead

Internal
Control
Vanes

Central Exploder Tube

Electric Fuse
For Warhead

Plywood Frame

Nitrogen Bottles

Aerial
Leads

Alcohol Filling Point

Conduit Carrying
Wires To Nose Of
Warhead

Oxygen Filling
Point

Front Joint Ring And
Strong Point For Transport

Stabilizing
Fins

Concertina
Connections

Double Walled Alcohol
Delivery Pipe To Pump

Pitch And Asimuth Gyros

Alcohol Pipes For
Subsidiary Cooling

Hydrogen Peroxide Tank

Nose Probably Fitted With Nose
Switch, Or Other Device For
Operating Warhead Fuse

Tubular Frames Holding
Turbine And Pump Assembly

Alcohol Inlet To Double Wall

Permanganate Tank
(Gas Generator Unit
Behind This Tank)

Electro Hydraulic
Servo Motors

Oxygen Distributor
From Pump

The V-2 was designed and tested at remote
Peenemünde, but after Allied bombing raids, production moved to widely separated sites. The main production center, Mittelwerk, was located underground in
the Harz Mountains in central Germany.

Building V-2s in quantity became a vast enterprise
in 1944. To meet production demands, Germany used
concentration camp prisoners to build the missiles
under unbearably harsh working conditions. Thousands
perished in the process. Near the war's end, nearly 700
V-2s were assembled each month in caverns near
Nordhausen.

Mittelwerk underground V-2 production site

The V-2 missile achieved Hitler's goal as a weapon of terror. From September 1944 until March 1945, 2,900 V-2s were fired against England, Belgium, and France from mobile launchers in Germany and its occupied territories. The V-2s were camouflaged to reduce their visibility to Allied bombers.

Traveling four times faster than sound and falling silently along their trajectory after engine shutdown, the missiles struck without warning. More than 1,100 V-2s hit southern England alone, causing an estimated 2,700 deaths and 6,500 injuries. Even more missiles were launched against the port city of Antwerp, in Belgium. Strikes were concentrated against population centers. Because the Germans could not pinpoint targets with precision, anyone within the surrounding area could be hit. V-2s killed a total of 7,000 people in Europe.

As Allied armies liberated Europe in early 1945, American, French, British, and Soviet military intelligence teams raced to capture information, matêriel, and personnel associated with German war technology. After surrendering to the U.S. Army, Wernher von Braun and other V-2 experts revealed their rockets' capabilities.

THIS IS A V-2 ROCKET
SEIZED BY U.S. ARMY ORDNANCE
TECHNICAL INTELLIGENCE TEAM N°1

V-2 in Washington

At the war's end, this V-2 was displayed in Washington, D.C., near 12th Street and Pennsylvania Avenue NW. It symbolized the end of the war—but also the new shape of possible conflicts to come.

The U.S. Army brought captured V-2 missile parts to White Sands Proving Ground, New Mexico, for its Project Hermes missile development program, managed by General Electric. Wernher von Braun and his team were housed at nearby Fort Bliss, Texas. They advised General Electric personnel in the reassembly, testing, firing, and evaluation of the missiles.

The first firings, in 1946, used all-German components. Later, modified American-made components were substituted to gain practical experience and to improve the basic missile design.

A General Electric technician inspects captured German V-2 engines at White Sands in late 1945.

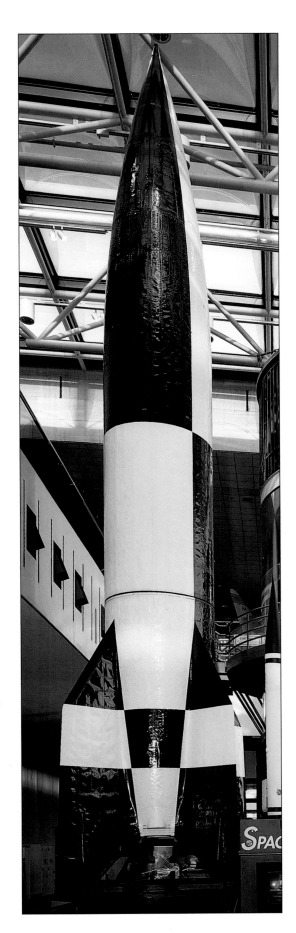

The Museum V-2

This V-2 missile, reconstructed by the U.S. Air Force using components from several missiles, was exhibited at the Air Force Technical Museum in Park Ridge, Illinois. The Smithsonian received it from the Air Force in 1954. Technicians spent more than 2,000 hours restoring it to its present appearance, which is patterned after the first successful test missile fired from Peenemünde in October 1942. These markings made the missile easily visible for accurate assessment of its flight performance. The missile's irregular surface reflects the condition of the original skin.

V-2 technical specifications:
    Length: 46 feet
    Weight: 28,000 pounds
    Range: 150–190 miles
    Maximum altitude: 50 miles for a 150-mile range
    Propellants: alcohol and liquid oxygen
    Manufacturer: German Army Ordnance

A Soviet copy of the German V-2 missile, called the R-1, was first launched in 1948.

The V-2 in the Soviet Union

The development of large-scale missile technology in the Soviet Union, as in the United States, was influenced by the German V-2. In 1947 the Soviets launched a V-2 assembled from German parts; a year later, they launched the first V-2 built in that country. This missile was called the R-1.

The Soviets went on to develop a variety of sounding rockets and missiles based on the V-2. They gradually increased engine thrust, enlarged the body, and integrated propellant tanks with the skin. These technical refinements increased the missile's range. The R-5, the last Soviet missile based on V-2 technology, had a range of 750 miles.

Catching Up with the V-2

While German V-2s were being used in wartime Europe, long-range missiles were still in the planning stages in the United States. Immediately after World War II, long-range ballistic missiles were not a top priority in the United States. Until a national policy was clarified in the 1950s, the U.S. armed services competed among themselves to develop missiles. This interservice rivalry led to early confusion and duplication of effort.

In addition to the Hermes and Corporal projects, the Army began work on larger rockets, including Redstone, Juno, and Jupiter, as well as the Nike anti-aircraft missile. The Navy developed the Aerobee and Viking sounding rockets, as well as smaller ship-to-air and ship-to-ship missiles. The Air Force inherited various industry proposals for long-range missiles from the army. One of these, the MX-774 program, eventually led to the Atlas intercontinental ballistic missile.

Courtesy of Los Alamos National Laboratory

The Ultimate Weapon

With the emergence of the Cold War, American and Soviet strategists confronted the same challenge—how to strike quickly at an enemy's heart. Both nations began to investigate means other than piloted aircraft to deliver bombs to distant targets.

At first they drew on German weapons technology from World War II, but the V-1 unpiloted flying bomb was too slow and vulnerable to enemy defenses, and the V-2 rocket was limited in both accuracy and range. Technological improvements gradually transformed the V-1 into the modern long-range cruise missile and the V-2 into the ICBM.

Traveling at least five times faster than sound (hypersonic speed) and independent of signals from the ground, the ICBM seemed to be the ultimate weapon.

By 1953, U.S. weapons designers had invented a way to make hydrogen bombs small and lightweight. This meant that the delivery system did not need to be as large as they had previously thought. The Bravo test, conducted in the South Pacific in 1954, confirmed the feasibility of the smaller new H-bomb design. The era of the ICBM was at hand.

SECURITY INFORMATION

SECRET

CONFIDENTIAL

TOP SECRET
Security Information

R W'006-4
Copy 2 of 16
Page 2 of 10

I. INTRODUCTION - SOME GENERAL REMARKS ON THE LONG-RANGE MISSILE PROGRAMS

1. The Committee's assignment has been limited to that of studying long-range intercontinental strategic missiles under development by the Air Force and making suitable recommendations for improving this program. Specific recommendations are made of changes for the improvement of the present Snark, Navaho, and Atlas programs.

2. Unusual urgency for a strategic missile capability can arise from one of two principal causes: a rapid strengthening of the Soviet defenses against our SAC manned bombers, or rapid progress by the Soviet in his own development of strategic missiles which would provide a compelling political and psychological reason for our own effort to proceed apace. The available intelligence data are insufficient to make possible a positive estimate of the progress being made by the Soviet in the development of intercontinental missiles. Evidence exists of an appreciation of this field on the part of the Soviet and of activity in some important phases of guided missiles which could have as an end objective the development by the Soviet of intercontinental missiles. While the evidence does not justify a conclusion that the Russians are ahead of us, it is also felt by the Committee that this possibility certainly cannot be ruled out.

3. Generally speaking, important aspects of the present long-range missile program consisting of the three projects, Snark, Navaho, and Atlas, are believed to be unsatisfactory. While specific recommendations for improving each of these programs are made in the following sections of this memorandum, certain weaknesses generally common to all programs are noted here.

    a. It is believed that all three missile systems have thoroughly out-of-date military specifications on target C. E. P.'s. This results from the very recent progress toward larger yield warheads which could hardly have been predicted when these specifications were originally established.

    b. The problem of reduction of base vulnerability needs much more careful study, particularly with respect to the influence on missile design that might be exerted by a better handling of this base vulnerability matter.

CONFIDENTIAL

SECURITY INFORMATION

SECRET

TOP SECRET
Security Information

The ICBM Decision

A top-secret report presented to the U.S. Air Force in early 1954 reassessed ballistic missiles in light of recent advances in nuclear weapons technology. Worried that the Soviets might be ahead of the United States in long-range ballistic missiles, the Strategic Missiles Evaluation Committee recommended that the Air Force treat missile development as "an extremely high priority."

Launching the Space Age

On October 4, 1957, a Soviet ICBM launched Sputnik—and the Space Age. This event startled the world, giving the impression that America was behind the Soviets in science and technology. Subsequent U.S. launch failures heightened that perception. The competition to build rockets now also became a competition to reach space.

From 1954 to 1957, Soviet rocket designer Sergei Korolëv headed development of the R-7, the world's first ICBM. Successfully flight-tested in August 1957, the R-7 missile was powerful enough to launch a nuclear warhead against the United States or to hurl a spacecraft into orbit.

Courtesy of Dmitri Kessel/Life Magazine ©Time, Inc.

Life, *October 21, 1957, "The Satellite: Why Reds Got It First."*

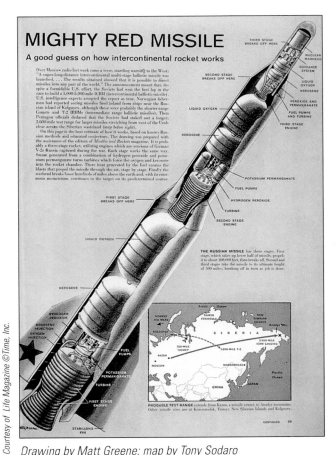

Courtesy of Life Magazine ©Time, Inc.

Drawing by Matt Greene; map by Tony Sodaro

In October 1957 the R-7 launched Sputnik, the world's first artificial satellite. In 1961 a modified R-7 launched the first manned spacecraft, Vostok, which carried cosmonaut Yuri Gagarin. A workhorse of the Soviet space program, the R-7 has launched many missions; refined versions are still in use today.

The successful test of the R-7 in August 1957 showed that the Soviets had the capability to launch a satellite into orbit. Yet the flight of Sputnik in October was an unexpected demonstration of Soviet technical prowess, and it had a great impact on public opinion in the United States. Americans were upset at being bested by their Cold War rivals and fearful that the Soviets could soon use missiles to launch a surprise nuclear attack.

Vanguard: America's Bid for Space

First the United States and then the Soviet Union announced intentions to place a scientific satellite into orbit as part of the 1957–1958 International Geophysical Year, a worldwide effort to study the Earth. The Army proposed to launch America's first satellite using a modified Redstone ballistic missile. Instead, Vanguard—a new vehicle descended from research rockets—was selected for technical reasons and also to emphasize the peaceful use of space.

After Sputnik's success, Vanguard's launchpad explosion on 6 December 1957 drew further attention to the Soviet lead in space.

The Vanguard was a three-stage launch vehicle developed for the Navy in 1957. It had liquid-propellant engines in the first two stages and a solid-propellant third stage. Vanguard's technical ancestors were the Navy's Viking and Aerobee sounding rockets.

Much of the internal hardware was removed when the Museum's Vanguard was prepared for display. It was transferred from the U.S. Naval Research Laboratory to the Smithsonian Institution in 1958 after being exhibited at Andrews Air Force Base in Maryland.

Vanguard technical specifications:
   Length: 70 feet, 9 inches
   Weight: 22,600 pounds
   Thrust: 27,000 pounds
   Propellants: (First stage) kerosene and liquid 3.201 oxygen; (Second stage) hydrazine and nitric acid; (Third stage) solid propellant
   Manufacturer: Glen L. Martin Company (prime);
   General Electric, Aerojet General, Thiokol (engines)

Jupiter-C: America Enters Space [see page 28]

In November 1957 the Soviet Union launched a second, much larger and heavier Sputnik. America's first success in space came on January 31, 1958, when Explorer 1 was launched aboard an Army Jupiter-C missile. In February a second U.S. attempt to launch a Vanguard satellite failed.

The American media and Congress demanded to know how the Soviets had beaten the United States into space. One response by the Eisenhower administration was to establish the National Aeronautics and Space Administration (NASA).

The Jupiter-C launched America's first satellite, Explorer 1, on January 31, 1958. Designed by the von Braun team and built by the Army, the Jupiter-C was a modified Redstone ballistic missile with a top stage designed for re-entry. Its liquid-propellant main-stage engine is derived from the Navaho missile program. The upper stages used solid propellants.

Both superpowers developed powerful missiles to deliver nuclear weapons. The U.S. deployed the Atlas, Titan, Minuteman, and Peacekeeper (MX) ICBMs on land, and the Polaris, Poseidon, and Trident submarine-launched ballistic missiles at sea.

With advances in technology, missiles became more accurate, capable of being launched on short notice, and able to carry multiple warheads. These advances, which made missiles more effective, also made them more inviting targets for attack.

America's first ICBMs, the Atlas and Titan missiles, also saw duty in the civilian space program as launch vehicles for spacecraft.

Atlas first carried the Mercury spacecraft bearing astronaut John Glenn into orbit in 1962. Later, Atlas missiles were mated to Able, Agena, and Centaur upper stages to create launch vehicles for a variety of American spacecraft.

Titan IIs launched all ten Gemini missions in 1965 and 1966. Later, Titans were fitted with upper stages and strap-on boosters to launch large planetary spacecraft and military satellites.

Vanguard *blows up on the pad*

Jupiter-C Explorer 1 launch, 1958

The Museum's Jupiter-C, transferred from the U.S. Army, is a full-scale model with a replica of Explorer 1 on top.

Jupiter-C technical specifications:

Length: 66 feet

Weight: 64,200 pounds

Thrust: 83,000 pounds

Propellants: hydrazine and liquid oxygen

Manufacturer: Chrysler (airframe), Rocketdyne (engine)

Atlas ICBM

America's First ICBM

The Atlas was deployed from 1959 to 1965. It was a "stage-and-a-half" liquid-propellant rocket with one main engine between two booster engines. The Museum's rocketry collection includes two Atlas launch vehicles, displayed at NASA visitor centers in Alabama and Florida.

The two-stage Titan ICBM became operational with the U.S. Strategic Air Command in 1962. It was deployed in "hardened" underground silos to protect it from attack. By 1965 an improved version, Titan II, was deployed; it was prefueled with storable propellants to reduce launch time.

Titan II ICBM

Launchable at a Minute's Notice

The silo-based, three-stage Minuteman missile became this country's standard ICBM. Unlike the first ICBMs, which used liquid propellants and were time-consuming to prepare for launch, the Minuteman was a solid-propellant missile ready for instant response.

Minuteman I was deployed in missile fields in the western and midwestern United States beginning in 1962. Each missile carried a single nuclear warhead. The original Minuteman was superseded by improved Minuteman II and Minuteman III missiles. Minuteman III could carry three independently targeted nuclear warheads. Some 550 Minuteman IIIs were deployed in the United States beginning in 1970.

Minuteman in silo

Operational Minuteman missiles are tested about three times a year. The Air Force selects a missile at random, removes its warheads, and ships the missile to Vandenberg Air Force Base in California. From there it is fired at a test target about 4,800 miles away, in the Marshall Islands of the South Pacific. These tests provide data on the accuracy and reliability of Minuteman components and the effects of aging on its solid propellant.

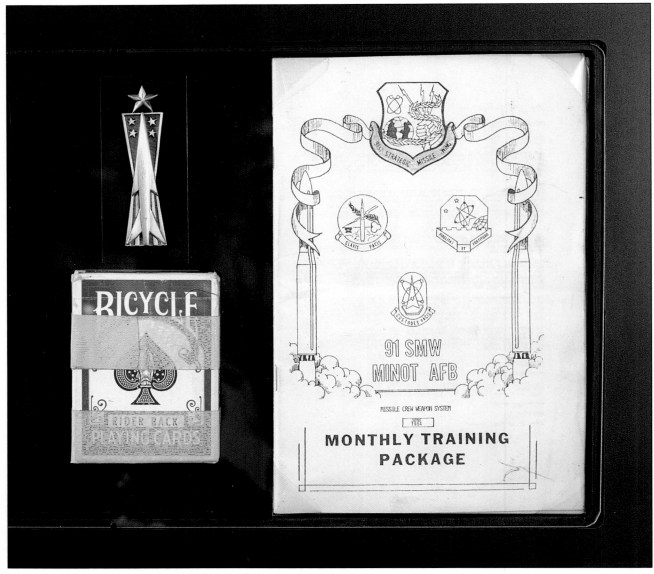

Technical order bag: gift of Capt. Robert F. Moore, U.S. Air Force. Pin: gift of U.S. Air Force

This Minuteman missileer's technical-order bag contained instructions for launch and related necessities, as well as a deck of cards for long hours on duty. The pin bears the Senior Missile Man insignia.

The Chip: Minuteman's Civilian Legacy

By 1970 the Minuteman had been modified so that it could be quickly retargeted. To accomplish this, in 1962 the Air Force chose a new and untested technology—the integrated circuit, or silicon chip—for an improved guidance computer. Minuteman production contracts helped to bring the chip from the laboratory into the consumer marketplace.

The chips shown here were among the early development models produced by Texas Instruments for the Minuteman program.

Courtesy of Los Alamos National Laboratory

Minuteman III MIRV warhead

Minuteman III

The Museum's Minuteman III missile, transferred from the U.S. Air Force, was prepared for training or static display. It contains neither propellant nor warheads.

Because the Minuteman is based underground and its lower casing has an ablative layer of cork—an organic material susceptible to fungus and decay—it has been treated with a fungicide, which accounts for the missile's green color.

Minuteman III technical specifications:
   Length: 60 feet
   Weight: 76,000 pounds
   Thrust: 170,000 pounds
   Propellant: solid
   Manufacturer (prime): Boeing Company

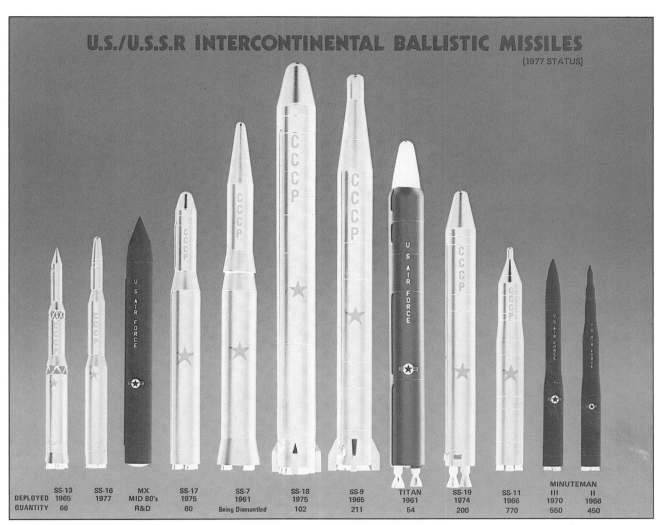

U.S./U.S.S.R INTERCONTINENTAL BALLISTIC MISSILES
(1977 STATUS)

	SS-13	SS-16	MX	SS-17	SS-7	SS-18	SS-9	TITAN	SS-19	SS-11	MINUTEMAN III	MINUTEMAN II
DEPLOYED	1965	1977	MID 80's	1975	1961	1975	1965	1961	1974	1966	1970	1966
QUANTITY	66		R&D	60	Being Dismantled	102	211	54	200	770	550	450

U.S. and Soviet ICBMs and SLBMs, c. 1977

Modern ICBMs

Today's ICBMs can reportedly strike within a few hundred feet of their targets after traveling thousands of miles through space. Such accuracy makes it possible to use a less powerful warhead and still be assured of destroying a target. A single missile can carry multiple warheads, each aimed at a different target. A Minuteman III could have three nuclear warheads, as shown here; the Peacekeeper MX had ten.

By the 1970s, multiple warheads and more accurate missiles increased concern that a first strike would destroy one side's missiles and bombers before they could be used to strike back. Both sides feared that a crisis might escalate into such a decisive attack. This threat caused further growth in the number of Soviet and American warheads and later led to arms limitation treaties. With the end of the Cold War, single warheads are replacing multiple ones, and nuclear arsenals are being reduced.

CHAPTER 2

Racing to the Moon

> "I believe that this nation should commit itself to achieving the goal, before this decade is out, of landing a man on the Moon and returning him safely to the Earth. No single space project. . . will be more exciting, or more impressive to mankind, or more important. . .and none will be so difficult or expensive to accomplish."
>
> —*President John F. Kennedy, 1961*

> "That's one small step for man, one giant leap for mankind."
>
> —*U.S. astronaut Neil Armstrong, on the moon, 1969*

At the start, there were no set rules for the space race. What was the goal? What would count as winning?

For Americans, President Kennedy's declaration focused the space race on a clear goal: landing a man on the moon before the Soviets did. The space race became a race to the moon.

For years the Soviets officially denied being in a race to the moon. Now there is ample evidence, including items displayed at this exhibition, that they competed to reach the moon first.

Publicity Versus Secrecy

The space race became a symbol of the broad ideological and political contest between two rival world powers. The way the two competitors organized to achieve their goals in space highlighted their basic differences.

The United States had separate civilian and military agencies, and only the military space programs were secret. Civilian space activities—especially the race to the moon—were openly publicized for the world to see.

In the Soviet Union, all space programs were integrated into a secretive military-industrial bureaucracy. Launches were not announced in advance, and only the successes were publicized.

During the early years of the space race, success was marked by headline-making firsts: the first satellite,

SHOULD YOU BUY A FOREIGN CAR?

POPULAR MECHANICS

WRITTEN SO YOU CAN UNDERSTAND IT

AUGUST 1959   35 CENTS

RACE TO THE MOON
Are the Russians Ahead?

Denver Post, *5 October 1957: "Russian Moon Circling Earth"*

first robotic spacecraft to the moon, first man in space, first woman in space, first spacewalk. To the dismay of the United States, each of these early feats was achieved by the Soviet Union. These events triggered a drive to catch up with—and surpass—the Soviets.

The Soviet Union stunned the world with the Sputnik launch on 4 October 1957. A shiny basketball-size sphere containing radio transmitters, Sputnik announced the beginning of the Space Age.

Coming just weeks after the Soviets' successful test launch of the first intercontinental ballistic missile, Sputnik signaled the Soviet Union's capability in rocketry and its potential to dominate space.

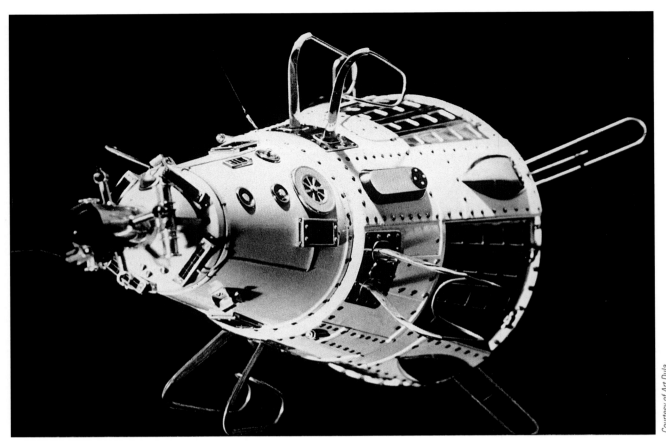

Sputnik 2

Sputnik Key

This metal arming key is the last remaining piece of the first artificial satellite to orbit the Earth. It prevented contact between the batteries and the transmitters until Sputnik was prepared for launch. A pin mounted on the launch vehicle served the same purpose until the satellite separated from the launcher in orbit. Only then did Sputnik begin to transmit the distinctive "beep, beep, beep" heard around the world.

Only a month after its "October surprise," the Soviet Union launched another satellite. Sputnik 2 was larger and carried a dog called Laika. Sputnik 2 demonstrated a growing Soviet advantage in launching heavy payloads and hinted that the Soviets might soon put a human in space.

From 1958 to 1961, six more Earth-orbiting Sputniks were successfully launched by the Soviet Union, all much larger than the first. These missions also improved reentry and recovery techniques for a human flight.

Courtesy of Emmet, Toni, and Tessa Stephenson

Sputnik 5 recovery card

Instructions for Sputnik Recovery

A card accompanied Sputnik 5, which carried the dogs Belka and Strelka into space in August 1960 on the Soviets' first successful capsule recovery mission. If the capsule landed outside the recovery zone, the card and related instructions directed anyone finding it to contact local officials immediately. The finder was also asked not to open the capsule but to set it upright, and to leave it exactly where it had landed.

Lent by Arthur M. Dula

Manufacturer: Experimental Design Bureau, OKB-1

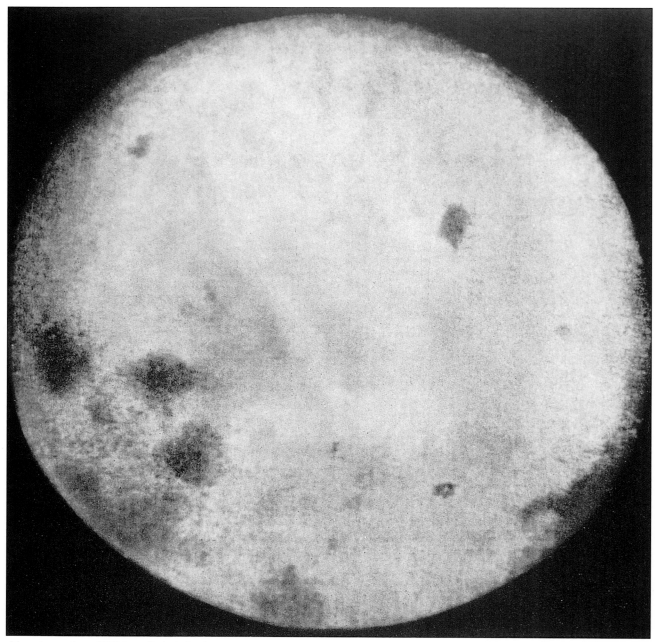

Moon image from 1959 Luna 3 photo book

Luna 3: First Flight around the Moon

On 4 October 1959, exactly two years after the first Sputnik launch, the Soviet Union sent the first spacecraft around the moon. Luna 3 recorded images of the moon's far side and broadcast them to Earth. A month earlier, after five unsuccessful attempts, the Soviet Luna 2 spacecraft had reached the moon.

This is one of the first images of the far side of the moon. It is reproduced from a book of Luna 3 images published in 1959 and presented by Sergei Korolëv, chief designer of the Soviet space program, to his wife. Korolëv inscribed it, "with good memories of the wonderful accomplishments of Soviet science."

Lent by The Perot Foundation

Ivan Ivanovich. Manufacturers: Moscow Prosthetic Appliances Works (mannequin); Zvezda (space suit)

Ivan Ivanovich: Test Flight Mannequin

This lifeless space traveler orbited the Earth on March 23, 1961, weeks before Yuri Gagarin's flight. His mission tested the Vostok spacecraft and SK-1 pressure suit, as well as tracking and recovery operations. The mannequin is named Ivan Ivanovich, the Russian equivalent of John Doe. Technicians were concerned because Ivan's features were so lifelike; they wrote "model" on the forehead, so that anyone finding the mannequin upon landing would not be confused.

Like subsequent Vostok cosmonauts, Ivan left the spacecraft in an ejection seat after reentering the atmosphere. He parachuted out of the seat and landed near the Ural Mountains city of Izevsk during a heavy snowstorm. Ivan has remained in his spacesuit for over thirty-five years.

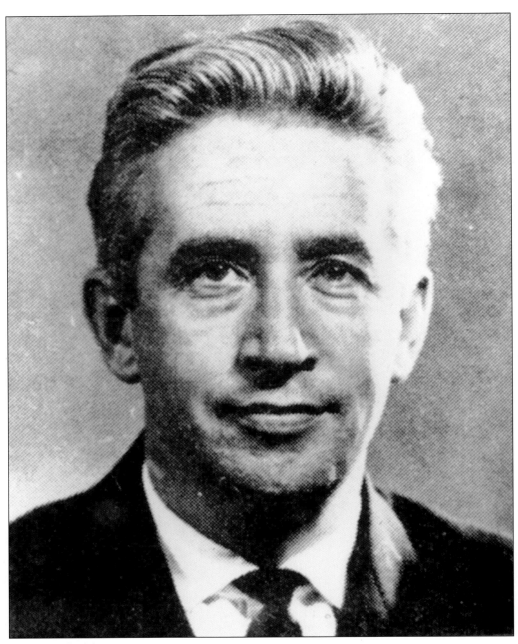

Konstantin Feoktistov

Soviet Secrets

Although the Soviet Union was achieving newsworthy firsts in space, very little was known in the West about its space program. Detailed information about missions and the identities of program managers and engineers were closely guarded state secrets. The notebooks of Konstantin Feoktistov, an engineer and cosmonaut whose importance was hidden for decades, contain rare, behind-the-scenes insights into the Soviet space program circa 1958–1959.

Pages from Feoktistov's notebook

Feoktistov's notebooks, which bear the censor's "Secret" stamps, were made public in 1989. In his notes, Feoktistov listed the research institutes and design bureaus responsible for each spacecraft component. These organizations were drawn from the Soviet Academy of Sciences, the Ministry of Defense, and other entities. Feoktistov referred to them by abbreviations and generic names.

Feoktistov's notebooks include sketches of manned spacecraft, an indication that shortly after Sputnik the Soviets were thinking of launching cosmonauts.

Yuri Gagarin: The First Man in Space

On 12 April 1961, the Soviets stunned the world again by sending a human into space. Cosmonaut Yuri Gagarin circled the Earth once in his Vostok spacecraft and returned safely. Gagarin's flight took place one month before American astronaut Alan Shepard's suborbital flight, and ten months before astronaut John Glenn became the first American to orbit the Earth. Once more, Gagarin's flight suggested that the Soviet Union was well ahead in the space race.

Gagarin delivered a speech to the Soviet State Commission on Space Flight two days before his historic flight. In it he thanked his colleagues "for trusting me to be the first to fly into space," adding that "I am glad, proud, happy—as any Soviet man would be." Gagarin also assured his audience that "I do not doubt the successful outcome of the flight."

Vostok Flight Plan

Feoktistov's draft checklist for the first man in space instructs Gagarin to follow certain procedures before launch, in orbit, and during descent. This copy was made from the original manuscript.

Vanguard

The United States had been planning to launch its first scientific satellite in late 1957. However, two launch attempts using the Navy's Vanguard rocket ended in disaster.

Public response to the Vanguard failures prompted national soul-searching in the United States. The media questioned why "Ivan" could accomplish things that "Johnny" could not.

This damaged Vanguard satellite was recovered after the December 6, 1957, launch attempt ended in an embarrassing explosion. Its "TV3" designation means "Test Vehicle Number 3."

London Daily Herald, *7 December 1957*

48

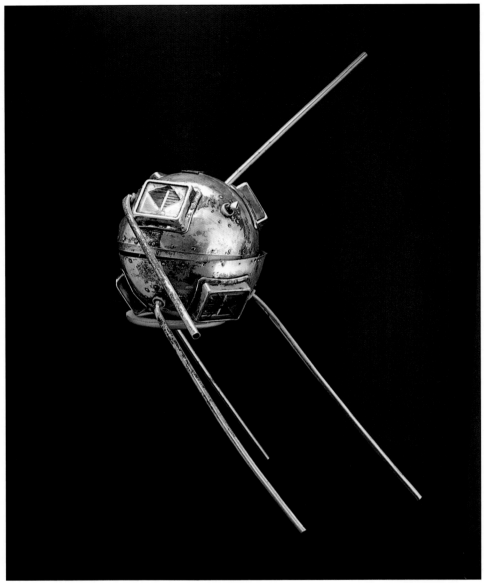

Vanguard TV-3 satellite

America's First Success

After the first Vanguard failure, the Army gained approval to attempt a satellite launch. On 31 January 1958 a modified Redstone missile, the Jupiter-C, lofted America's first satellite, Explorer 1, into space. In March the Navy's Vanguard succeeded in its third attempt to launch a satellite.

Although still behind, America had rallied after its initial stumble and was now in the space race.

Immediately after Gagarin's flight, President Kennedy wanted to know what the United States could do in space to take the lead from the Soviets. Vice President Lyndon Johnson polled leaders in NASA, industry, and the military. He reported that "with a strong effort," the United States "could conceivably" beat the Soviets in sending a man around the moon or landing a man on the moon. As neither nation had a rocket powerful enough for such a mission, the race to the moon was a contest that the United States would not start at a disadvantage.

On May 25, 1961, when President Kennedy announced the goal of landing a man on the moon, the total time spent in space by an American was barely fifteen minutes.

Hundreds of thousands of people in many

Huntsville Times, *1 February 1958*

organizations worked in the American and Soviet space programs. There were dreamers, brilliant engineers, and talented managers on both sides, but the careers of two individuals with important technical and managerial roles illustrate some of the differences. The American was a well-known personality in the highly publicized U.S. civilian space program. The Soviet was officially, anonymously known as the Chief Designer; he was not publicly identified until after his death.

Shortly after NASA was created, the U.S. Army Ballistic Missile Agency research group led by Wernher von Braun was transferred to the civilian space agency, and in 1960 it became the core of NASA's Marshall Space Flight Center in Huntsville, Alabama.

Von Braun served until 1970 as the first director of the Marshall Center, which was responsible for developing rocket engines and launch vehicles, including the American "Moon rocket," the giant Saturn V.

An avid proponent of space exploration, von Braun collaborated on a series of popular 1950s magazine articles and television shows depicting future space travel.

Time, *17 February 1958*

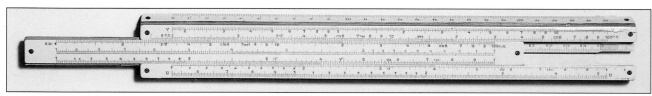

Manager and public figure Wernher von Braun stayed close to the work of engineering and design. He used this slide rule, made by Albert Nestler A.G. of Germany, for calculations.

Courtesy of RSC Energia

Sergei Korolëv

Sergei Korolëv

In the 1930s, Russian engineer and aviator Sergei P. Korolëv (1907–1966) headed GIRD, a Moscow-based group of rocket enthusiasts that built and tested the first liquid-propellant rockets in the Soviet Union.

After World War II, Korolëv was appointed to head a Soviet missile-development design bureau. In 1957 his bureau launched the R-7—the first operational ICBM, which was used to propel Sputniks into Earth orbit and Luna spacecraft to the moon.

Korolëv's work defined the Soviet school of rocket and spacecraft design, including the Vostok and Soyuz manned spacecraft, ballistic missiles and scientific rockets, the Zenit reconnaissance satellite, Molniya communications satellites, and manned lunar spacecraft. Korolëv's design bureau has evolved into a Russian business organization known today as the Energia Rocket and Space Corporation, or RSC Energia.

Korolëv used this German slide rule (like von Braun's, a Nestler) to make quick calculations. To his colleagues, Korolëv's slide rule was a magician's wand. Today's engineers and scientists use pocket calculators and computers for the same purposes.

Lent by The Perot Foundation

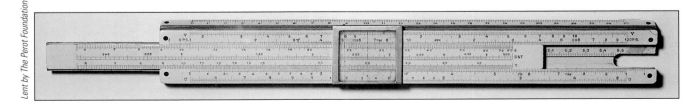

Courtesy of The Perot Foundation

This informal autographed portrait of the Vostok cosmonauts dates from the early sixties. From left to right: Pavel Popovich, Yuri Gagarin, Valentina Tereshkova, Valery Bykovsky, Andrian Nikolayev, and Gherman Titov. All but Tereshkova were military pilots.

National Heroes

Russia's cosmonauts and America's astronauts became the most visible symbols of the space race. These young space pilots were celebrated as national heroes, and their flights were widely heralded around the world.

Vostok and Voskhod

The Vostok and Voskhod missions of 1961 to 1965 continued the series of Soviet firsts in space. In each of six missions from 1961 to 1963, a Vostok ("East") space-craft carried a cosmonaut into Earth orbit in successively longer flights.

The Vostok spacecraft then was modified to hold two or three cosmonauts and renamed Voskhod ("Sunrise"). Three cosmonauts orbited aboard Voskhod 1 for a day in October 1964, five months before the first U.S. two-man Gemini mission. In March 1965, Voskhod 2 achieved another space spectacular, the first spacewalk, when cosmonaut Aleksei Leonov ventured outside his orbiting spacecraft.

The Mercury 7 astronauts were selected and introduced to the public in 1959. All were military test pilots. From left to right: (front row) Walter Schirra, Donald Slayton, John Glenn, and Scott Carpenter; (back row) Alan Shepard, Virgil "Gus" Grissom, and Gordon Cooper.

Courtesy of Art Dula

Vostok spacecraft

Soviet Firsts in Vostok

1961: First man in space, Yuri Gagarin's one-orbit flight (Vostok 1)

1961: First full day in orbit, Gherman Titov (Vostok 2)

1962: First two-spacecraft mission (Vostoks 3 and 4)

1963: First long-duration mission, five days in orbit (Vostok 5)

1963: First woman in space, Valentina Tereshkova (Vostok 6)

This photograph was made from motion picture footage of cosmonaut Valentina Tereshkova during her 1963 flight on Vostok 6.

An Engineer-Cosmonaut

As a reward for redesigning the Vostok capsule to accommodate more than one cosmonaut, Soviet spacecraft designer Konstantin Feoktistov was selected to join the crew of the 1964 Voskhod mission.

To fit three cosmonauts into the crowded spacecraft, Soviet designers made a risky decision: the cosmonauts would not sport the bulky pressure suits usually worn as a precaution against loss of pressure in the capsule. Konstantin Feoktistov wore this wool two-piece flight suit (p. 56) on Voskhod 1 instead of a pressure suit.

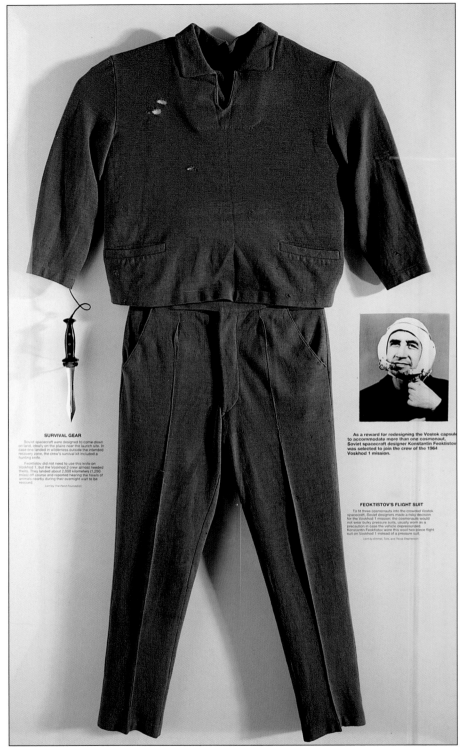

SURVIVAL GEAR

Soviet spacecraft were designed to come down on land, ideally on the plains near the launch site. In case one landed in wilderness outside the intended recovery zone, the crew's survival kit included a hunting knife.

Feoktistov did not need to use this knife on Voskhod 1, but the Voskhod 2 crew almost needed theirs. They landed about 2,000 kilometers (1,250 miles) off course and reported hearing the howls of animals nearby during their overnight wait to be rescued.

Lent by The Perot Foundation

As a reward for redesigning the Vostok capsule to accommodate more than one cosmonaut, Soviet spacecraft designer Konstantin Feoktistov was selected to join the crew of the 1964 Voskhod 1 mission.

FEOKTISTOV'S FLIGHT SUIT

To fit three cosmonauts into the crowded Voskhod spacecraft, Soviet designers made a risky decision for the Voskhod 1 mission: the cosmonauts would not wear bulky pressure suits, usually worn as a precaution in case the vehicle depressurized. Konstantin Feoktistov wore this wool two-piece flight suit on Voskhod 1 instead of a pressure suit.

Lent by Emmet, Toni, and Tessa Stephenson

Suit lent by Emmet, Toni, and Tessa Stephenson; knife lent by The Perot Foundation

Soviet spacecraft were designed to come down on land—ideally on the plains near the launch site. In case one landed in wilderness outside the intended recovery zone, the crew's survival kit included a hunting knife.

Feoktistov did not need to use this knife on Voskhod 1, but the Voskhod 2 crew almost needed theirs. They landed 1,250 miles off course and reported hearing the howls of animals nearby during their overnight wait to be rescued.

Courtesy of Zvezda

The First Spacewalk

On 18 March 1965, Aleksei Leonov became the first person to venture outside an orbiting spacecraft. He was secured only by an umbilical cord attached to the life-support systems of Voskhod 2. Leonov spent twenty minutes outside in the vacuum of space.

To allow a cosmonaut to leave the pressurized spacecraft, Soviet engineers designed a flexible airlock attachment. Once inflated in orbit, the airlock became a tunnel for exiting and reentering the spacecraft, which remained sealed. Aleksei Leonov trained for his spacewalk in this airlock. For American spacewalks, the spacecraft was completely depressurized and the hatch opened directly to space.

Manufacturer: Zvezda.

The first man to walk in space almost became the first to die there. During the March 1965 flight of Voshkod 2, cosmonaut Aleksei Leonov eased headfirst into the airlock, and then into space. Leonov later recalled that he felt like "a bird, with wings and able to fly." But as the cosmonaut attempted to reenter the airlock, his rapture turned to shock. In the vacuum of space, internal air pressure had inflated the size of the suit and increased its rigidity. The suit was too large and inflexible to fit into the airlock. Racing through its orbit, the spacecraft would soon plunge into complete darkness, making entry into the airlock nearly impossible. Leonov tried an untested, risky solution: venting some of the suit's precious air into empty space. It worked; Leonov struggled back into the airlock and the safety of the Voshkod capsule. He later received a reprimand from his superiors. "They were dismayed that I had made the decision to release pressure from the suit. We were not supposed to make any decisions without consulting Ground Control."

Leonov devised a "bracelet" to keep a set of colored pencils handy, so he could sketch the view from space.

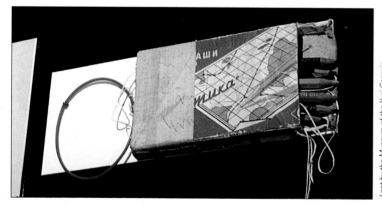

Courtesy of NASA

John Glenn, first American in orbit

Mercury and Gemini

Early U.S. manned spaceflights were spectacularly successful. In May 1961, American astronaut Alan Shepard went briefly into space (though not into orbit) on the Mercury 3 mission. In February 1962, John Glenn spent five hours in orbit on Mercury 6. In June 1965, Gemini IV astronaut Edward White made the first U.S. spacewalk.

Although it seemed that the United States still lagged behind the Soviet Union in space, the United States was following a methodical step-by-step program in which each mission built upon and extended the previous ones. The Mercury and Gemini missions carefully prepared the way for the Apollo lunar missions.

Mercury spacecraft

Courtesy of NASA

The one-man Mercury missions developed hardware for safe spaceflight and return to Earth, and began to show how human beings would fare in space. From 1961 to 1963, the United States flew many test flights and six manned Mercury missions.

After Mercury, NASA introduced Gemini, an enlarged, redesigned spacecraft for two astronauts.

Ten manned Gemini missions from 1964 to 1966 improved techniques of spacecraft control, rendezvous and docking, and extravehicular activity (spacewalking). One Gemini mission spent a recordbreaking two weeks in space—time enough for a future crew to go to the moon, explore, and return.

Gemini spacecraft

Ed White's EVA

Astronaut Edward White's spacewalk during the 1965 flight of Gemini 4 provided one of the enduring images of the space race. Tethered to the capsule, White maneuvered and tumbled alone in space for about twenty minutes. Television captured White's small figure floating against the backdrop of Earth and space. Leonov may have performed the first spacewalk, but live television and the American policy of openness made viewers everywhere participants in the nation's effort to catch the Soviets.

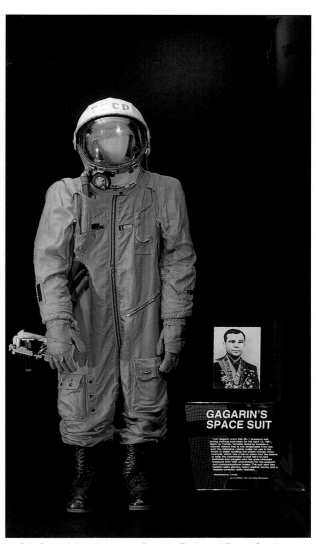

Glenn's suit transferred from NASA; Gagarin's suit lent by Emmet, Toni, and Tessa Stephenson

United States and Soviet Space Suits

[Left] John Glenn wore this space suit on February 20, 1962, when he became the first American to orbit the Earth. Like the Gagarin suit next door, its design was adapted from high-altitude pressure suits worn by aircraft pilots. Glenn's suit was a lightweight multilayered garment with an aluminized nylon cover layer. Thirteen zippers, plus custom-fitted gloves, boots, and helmet, assured a snug fit. This suit has become fragile with age. A mannequin, smaller than the astronaut, has been inserted to preserve its shape.

[Right] Yuri Gagarin wore this SK-1 pressure suit during training exercises for his 12 April 1961 flight on Vostok. Manufactured by Zvezda, its notable features include a visored helmet that is not detachable from the suit; an inflatable rubber collar for water landing; a bright orange nylon oversuit with a mirror sewn into the sleeve to help locate hard-to-see switches and gauges; and a gray-checked pressure liner with connectors for life support and communications hoses. The suit also has leather-palmed gloves, heavy leather boots, and a leather-covered radio headset.

Courtesy of RSC Energia

Vasily Mishin

A Soviet Moonshot

From 1958 through 1976, the Soviet Union sent automated explorers that circled, landed on, and roamed about the moon. Three robotic craft even gathered samples of lunar soil and brought them to Earth. Yet the U.S.S.R. never announced its intent to land a cosmonaut on the moon.

With the end of the Cold War, Soviet plans to send men to the moon have come to light. Newly released diaries, technical documents, and space hardware offer glimpses of the Soviet Union's ambitious manned lunar program. A prototype lunar space suit shows that the Soviets were serious about landing on the moon.

Courtesy of The Perot Foundation

A page from the Mishin Diaries, 1965

The Mishin Diaries

Rocket engineer Vasily Mishin served as deputy to Sergei Korolëv in the Experimental Design Bureau, working closely with him on many space projects. When Korolëv died in 1966, Mishin became Chief Designer and inherited responsibility for the Soviet manned lunar program.

From 1960 to 1974, Mishin kept private diaries detailing the day-to-day workings and decisions of the Soviet space program.

As early as 1960, Mishin recorded that Korolëv was very disappointed by debates and government delays in adopting a master plan for long-term scientific space exploration, including human flights to the moon and Mars. The Soviet decision in favor of a manned lunar program came after the United States had set the ultimate goal in the space race—landing a man on the moon.

In a 1965 entry, Mishin summarized upcoming Soviet space activities in which the design bureau would play a leading role. He mentioned military satellites, space stations, space planes, and various activities on the moon. He also listed many of the items and operations needed for a manned landing on the moon, including special tools, maps, and space suits.

In 1967, Mishin recorded the space achievements planned to commemorate the fiftieth anniversary of the Bolshevik Revolution. A manned flight around the moon and a test of the N-1 moon rocket are on the list.

Soviet Lunar Programs

The Soviets had several programs to explore the moon:

Luna (a variety of automated orbiters, landers, and soil sample return capsules, 1959–1976);

L-1/Zond (automated circumlunar missions, trial runs for a manned "loop-around-the-Moon" flight, 1965–1970);

Soyuz and Kosmos (manned and automated missions in Earth orbit to test lunar spacecraft and maneuvers, 1966–1969);

Lunokhod (automated lunar rovers, 1970–1973);

L3 (a never-executed "Man-on-the-Moon" landing, originally scheduled for late 1968).

Only twenty of about sixty Soviet launches of lunar probes from 1959 through 1976 were successful.

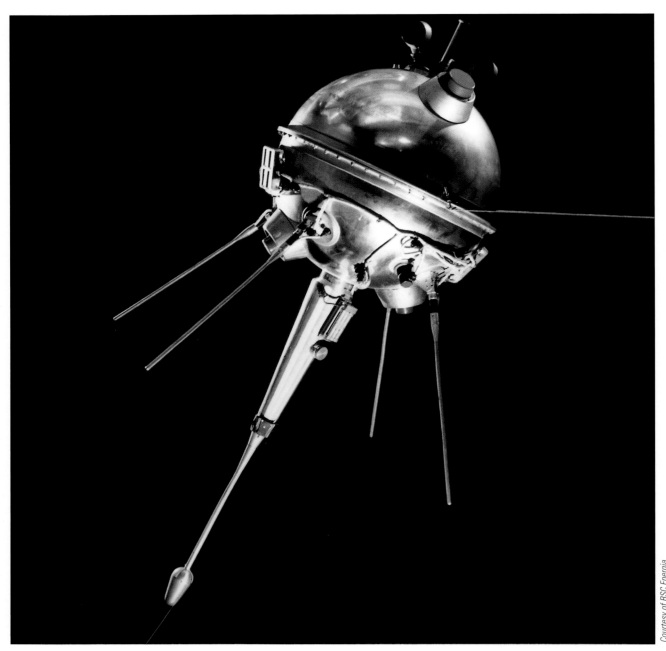

Courtesy of RSC Energia

Luna 2

Korolëv's Design Bureau began work in 1965 on the L-1, a manned spacecraft intended to carry two cosmonauts on a single loop around the moon. Because of repeated equipment failures, L-1 never flew with a crew. However, unmanned L-1 spacecraft flew to the moon five times under the name Zond ("Probe") from 1968 to 1970 to test the spacecraft and maneuvers necessary for a manned lunar mission. In September 1968, Zond 5 became the first spacecraft to loop around the moon and return to Earth.

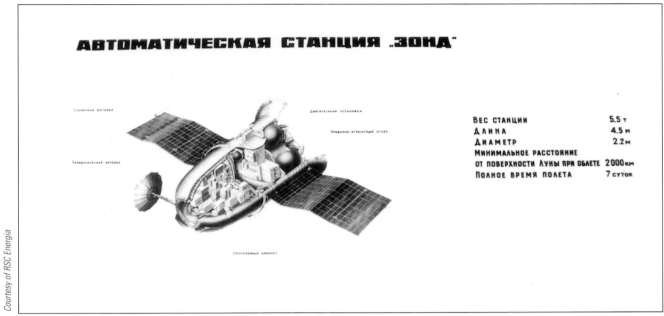

L-1 spacecraft

Two Luna missions in 1970 and 1973 included a robotic rover, Lunokhod, that roamed around the landing site. Lunokhods were equipped to take photographs and to analyze rock and soil samples—the same kinds of tasks performed by astronauts on the moon. Although these Soviet robotic explorers were successful, they were overshadowed by the American-manned explorations.

Lunokhod

Korolëv also began designing spacecraft for a lunar landing mission, and hardware was built under Mishin's direction. The manned lunar landing program called L-3 included an orbiter and lander. The prototype lunar lander was successfully tested in Earth orbit, without a crew, three times in 1970 and 1971 under the name Kosmos.

The Soviet lunar lander was half as large and one-third the weight of the U.S. Apollo lunar module. It was intended to carry one cosmonaut to the surface of the moon while the other stayed in lunar orbit. The program was canceled without a manned flight after repeated test failures of the launch vehicle.

L-3 lander

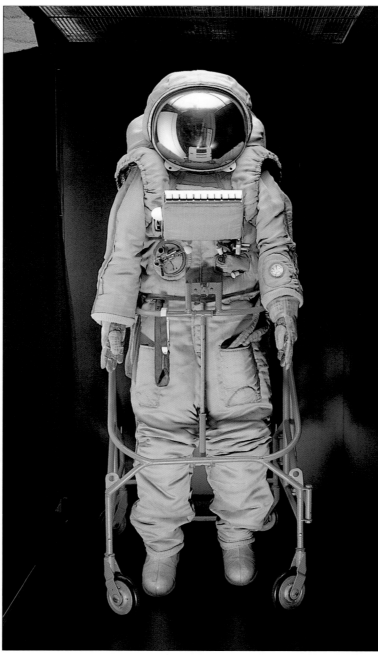

A Soviet Moon Suit

The Soviets developed this space suit for use by a cosmonaut on the moon. Called Krechet ("Golden Falcon"), it differs from the Apollo space suit in several ways. The backpack life support unit is hinged like a door, allowing the cosmonaut to step into the suit; although the arms and legs are flexible, the torso of the Krechet suit is a semirigid shell; the control panel on the chest folds up out of the way when not in use; and the boots are made of flexible leather.

Like the Apollo helmet, the Krechet helmet has a gold-coated outer visor for protection from bright sunlight. The life support backpacks are also similar, containing systems to provide oxygen, suit pressure, temperature and humidity control, and communications.

A similar space suit is used by cosmonauts working outside the Russian space station Mir.

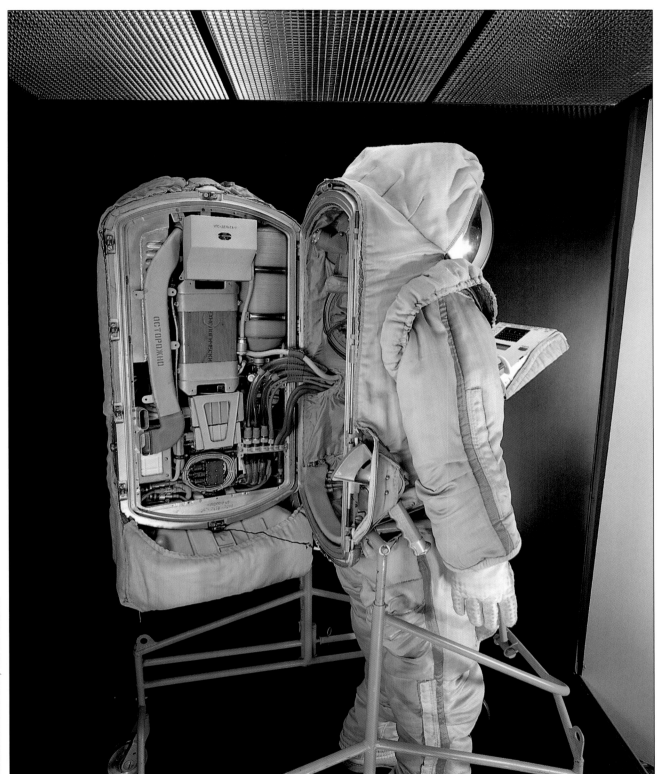

Transferred from NASA

Manufacturer: ILC Industries, Inc.

Apollo Lunar Suit

This space suit was worn on the moon by Apollo 15 astronaut David Scott in 1971. Moondust is still visible on the legs and boots.

The space suits had to meet all the astronauts' life support needs. Backpacks (left on the moon) provided oxygen, temperature and humidity control, suit pressure, and power for communications and data display systems. This suit, made of twenty-two layers of several materials plus a three-layer undergarment, also protected them against extreme lunar temperatures and micrometeoroids.

Because the lunar explorers had to walk over uneven terrain, collect rock and soil samples, and set up equipment, the suit was sturdy but gave them adequate mobility. Each astronaut's suit was custom-fitted.

All Dressed Up But No Way to Go

Soviet cosmonauts had a lunar orbiter, a lunar lander, and a space suit for the moon. Why didn't they go? The crucial missing piece was a rocket powerful and reliable enough to send a manned spacecraft to the moon.

When the space race began, no one had a rocket powerful enough to send a man to the moon and back. Both Americans and Soviets had to develop a super-booster, or moon rocket. The United States succeeded with the mighty Saturn V. The Soviets' N-1 moon rocket never made it into space.

Bottom view of N-1 first stage

The N-1 Rocket

In the early 1960s, Korolëv's design bureau began work on a multipurpose heavy-lift rocket—the N-1. In 1964 it was approved for redesign and use in the manned lunar program.

Begun under Korolëv and tested under Mishin, the N-1 rocket suffered from critical technical problems that doomed Soviet efforts to land a man on the moon by 1970. All four unmanned N-1 flight tests ended in failure. The N-1 effort was canceled in 1974, and the

Soviet manned lunar program passed into oblivion.

In the first launch attempt in February 1969, an engine fire caused the N-1 rocket to shut down and crash a minute after liftoff.

The second test, in July 1969, was a greater disaster. The rocket shut down seconds after liftoff, fell onto the launchpad, and exploded. The top photograph on the opposite page shows the test rocket just before its destruction. The accident destroyed the launch site and

Liftoff, second N-1 launch attempt

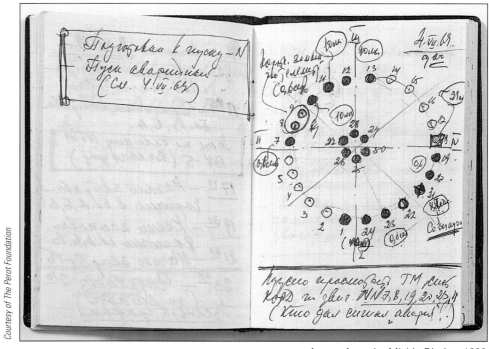

A page from the Mishin Diaries, 1969

BAIKONUR COSMODROME, USSR: 1969

Launchpads for 1960s Race to the Moon
24 September 1968

Post Explosion
3 August 1969

These extensive intercontinental ballistic missile (ICBM) and space launch facilities in Kazakhstan are known in the West as Tyuratam Missile Test Center. The complex was first seen on U-2 aircraft photography in 1957, when there was only one launchpad. By 1972, the facilities had grown to 78 missile/space launch pads and silos spread across a 40-nautical-mile rangehead. Soviet liquid-fueled ICBM systems were tested here, and all of the Soviet manned space launches were from the original launchpad.

During the 1960s, the Soviets entered into a huge program to put a man on the moon ahead of the US. The launch facilities for this program, known as launch complex J, took over six years to construct. After several failures—one an on-pad explosion that heavily damaged one launchpad as shown on this graphic—the program was terminated. Only after the dissolution of the USSR have the Russians released information about this effort.

Courtesy of the Central Intelligence Agency and the National Reconnaissance Office

any hope that the Soviets could reach the moon ahead of the United States.

Three weeks later, the Apollo 11 crew landed on the moon.

Mishin's diaries include a sketch (previous page) of the first-stage engines, showing how the automatic engine control system failed in the devastating July 1969 test launch.

A U.S. Corona reconnaissance satellite took this picture of the demolished Soviet launch site after the July 1969 N-1 launch attempt.

How did the Soviet Union lose its early lead in the space race and fail to send cosmonauts to the moon? Since all of the N-1 failures occurred in the thirty-engine first stage, Western analysts have speculated that the Soviets were unable to develop a sufficiently

powerful and reliable rocket engine in time to beat the United States to the moon.

Recently released diaries and memoirs of Soviet participants suggest that other problems also contributed to the N-1 failures. For example, in 1966 Mishin noted several deficiencies in the Soviet space program, including problems with the supply of hardware components, the absence of a national space agency, the low priority of the manned lunar program, and the lack of a long-term master plan for space exploration.

The N-1 was a three-stage giant, with thirty first-stage engines, eight second-stage engines, and four third-stage engines. There were also two single-engine stages for the spacecraft payload. The airframe was open between stages to vent exhaust because each upper stage ignited before the lower one was jettisoned. Propellants for all stages were kerosene and liquid oxygen.

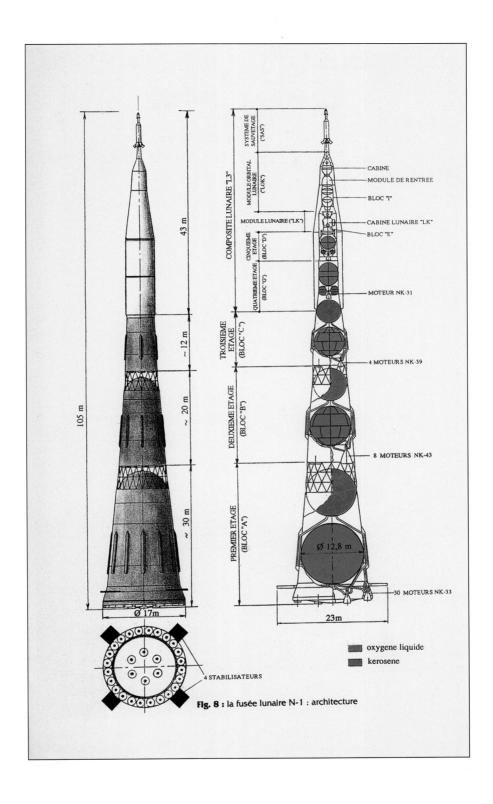

Fig. 8 : la fusée lunaire N-1 : architecture

Ten N-1s were built. Four were destroyed in failed test launches, and the others were dismantled when the program was canceled in 1974.

N-1 technical specifications:
Size: 347 feet
Thrust: 9,900,000 pounds
Payload to orbit: 209,000 pounds
Payload to moon: 66,000 pounds
Manufacturer: Experimental Design Bureau OKB-1

Courtesy of NASA

A Saturn V launch

Saturn V: America's Moon Rocket

Saturn V, developed at NASA's Marshall Space Flight Center under the direction of Wernher von Braun, was the largest in a family of liquid-propelled rockets that solved the problem of getting to the moon. Thirty-two Saturns were launched; none failed.

Saturn V was flight-tested twice without a crew. The first manned Saturn V sent the Apollo 8 astronauts into orbit around the moon in December 1968. After two more missions to test the lunar landing vehicle, in July 1969 a Saturn V launched the crew of Apollo 11 to the first manned landing on the moon.

Nearly forty stories tall, the three-stage Saturn V was the largest, most powerful rocket ever launched. With a cluster of five powerful engines in each of the first two stages and using high-performance liquid hydrogen fuel for the upper stages, the Saturn V was one of the great feats of twentieth-century engineering. Inside, the rocket contained three million parts in a labyrinth of fuel lines, pumps, gauges, sensors, circuits, and switches—each of which had to function reliably.

Fifteen Saturn Vs were built. The Museum's collection includes three Saturn Vs exhibited at NASA visitor centers in Alabama, Florida, and Texas.

Saturn V technical specifications:
  Size: 363 feet
  Payload to orbit: 285,000 pounds
  Payload to moon: 107,000 pounds
  Manufacturer: Boeing Company (prime)

Courtesy of NASA

Saturn's aft end

First stage: five F-1 engines

Propellants: RP-1 (kerosene) and liquid oxygen

Total thrust: 7,500,000 pounds

Manufacturer: Rocketdyne

Second stage: five J-2 engines

Propellants: liquid hydrogen and liquid oxygen

Total thrust: 1,250,000 pounds

Manufacturer: Rocketdyne

Third stage: one J-2 engine

Thrust: 250,000 pounds

80 The pace of the race to the moon quickened in late 1968. The following timeline tells the events that occurred between September 1968 and July 1969.

1968

September

Soviet Zond 5 unmanned test flight loops around moon and returns to Earth.

October

U.S. Apollo 7 manned test flight of command and service modules in Earth orbit.
Unmanned Zond 6 circumlunar flight.

December

Soviet manned flight to moon canceled after October Zond problems.
Apollo 8 crew orbits moon and returns safely.

1969

February

Soviet attempt to launch N-1 moon rocket fails.

March

Apollo 9 test of lunar module in Earth orbit.

May

Apollo 10 test flight of lunar module, with descent from lunar orbit to low altitude above moon.

July

Second Soviet N-1 launch failure.
Launch of Luna 15 lander for robotic collection and return of moon rocks (crashed).
Apollo 11 crew succeeds in first landing on the moon.

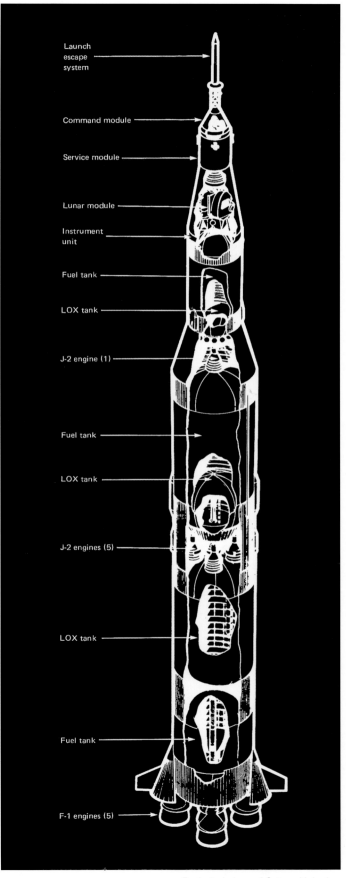

Exploded view of Saturn V design

The end of the moon race appeared imminent with the successful completion of the Apollo 8 and Apollo 10 missions.

In a suspenseful first foray, the crew of Apollo 8 looped around the moon in December 1968. They were the first people to see "Earthrise." Five months later, the Apollo 10 crew went into lunar orbit and tested the lunar module in a partial descent to the moon.

These missions built confidence that the United States was ready to proceed with the lunar landing. The big question was what the Soviets were planning to do.

Courtesy of NASA

Earthrise, seen from Apollo 8

When it became evident after the second N-1 rocket launch failure that the Soviet Union could not send a man to the moon ahead of the Americans, the Soviets attempted to obtain the first lunar rock and soil samples, sending a robot instead of a cosmonaut.

Luna 15, an automated sample return craft, was launched to the moon two days before Apollo 11. It crash-landed there shortly after U.S. astronauts Neil Armstrong and Buzz Aldrin first stepped onto the moon. If the Luna 15 lander had not crashed, it would have returned to Earth with lunar soil just hours ahead of the Apollo 11 crew.

The Moon Race Ends

On 20 July 1969, as millions around the world watched on television, two Americans stepped onto another world for the first time. The United States successfully landed men on the moon and returned them safely, fulfilling President Kennedy's vision and meeting the goal that inspired manned spaceflight during the 1960s.

The lunar landing was celebrated as an epic technological achievement and a triumph of the human spirit. In the span of a lifetime, humans had made a giant leap from the Wright brothers' first powered flight on Earth to the first steps on another planet.

Time, July 25, 1969

Courtesy of NASA

Triumph and Tragedy

Spaceflight is risky. The exploration of space has not been accomplished without loss of life.

In January 1967, during training for the first Apollo mission, astronauts Virgil "Gus" Grissom, Edward White, and Roger Chaffee died when a flash fire erupted in their spacecraft on the launchpad. U.S. manned flights were halted for almost two years while the Apollo spacecraft was redesigned.

In April 1967 the flight of Soyuz 1 ended in tragedy when the capsule's descent parachute failed to open. Cosmonaut Vladimir Komarov died in the crash landing, and the next manned Soyuz flight was delayed for eighteen months.

Courtesy of NASA

A Soyuz Sorrow

A poignant reminder of the risks of spaceflight, the doll at right was autographed by cosmonaut Viktor Patsayev just before the ill-fated *Soyuz 11* flight in 1971. After three successful weeks aboard the *Salyut 1* space station, Patsayev and his comrades Georgi Dobrovolsky and Vladislav Volkov died during descent when their capsule depressurized. To fit into the small spacecraft, the three-man crew had to forego wearing space suits. Patsayev had postdated his signature to the day after his planned return.

Among the items left on the moon by the Apollo 15 crew was a small memorial (above) to the astronauts and cosmonauts who had lost their lives in the quest to explore space.

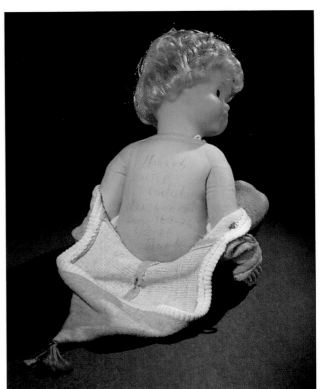

Lent by Emmet, Toni, and Tessa Stephenson

Apollo 17

After the Moon Race

When the race to the moon ended, the Soviet and American manned spaceflight programs moved in other directions. In the United States, many expected the Apollo missions to begin an era in which humans would move into space—to bases on the moon and space stations in Earth orbit, or perhaps to Mars. Others questioned whether costly manned spaceflight should continue now that the race was won.

Courtesy of NASA

For the Soviets, the competition with the United States did not end when they began to pursue longer-term goals, such as establishing a permanent presence in space with a series of Earth-orbiting space stations. The U.S.S.R. also sent automated probes to explore the surfaces of Venus and Mars.

The American space effort shifted to other programs as the last lunar mission, Apollo 17, was completed in 1972. In 1973 and early 1974, U.S. astronauts occupied Skylab, an experimental space station adapted from Apollo hardware. The focus in the 1970s was on developing a new vehicle—a reusable Space Shuttle—for future human missions in Earth orbit. The United States also sent robotic explorers to Mars and the outer planets.

CHAPTER 3

Secret Eyes in Space

90

"We've spent between thirty-five and forty billion dollars on space . . . but if nothing else had come from that program except the knowledge that we get from our satellite photography, it would be worth ten times to us what the whole program has cost. Because tonight I know how many missiles the enemy has and . . . our guesses were way off. And we were doing things that we didn't need to do. We were building things that we didn't need to build. We were harboring fears that we didn't need to have."

—*President Lyndon B. Johnson, 1967*

Photography from spy satellites is a significant legacy of the space race and the Cold War. Reconnaissance was one of the first priorities of spaceflight.

From 1960 to 1972, in a reconnaissance project code-named Corona, the United States routinely photographed the Soviet Union from space. In difficulty, the Corona project rivaled the public drama of sending men to the moon, but its successes are generally unknown: spying from space is top secret.

Corona was mostly a response to fear of nuclear attack by an intensely secretive Soviet Union. America's leaders faced an urgent question: What were the Soviets actually doing behind the Iron Curtain? Corona provided vital answers.

Reconnaissance and Space

In the mid-1950s, President Dwight Eisenhower was worried about the possibility of a surprise nuclear attack by the Soviet Union. The United States had two choices to ease these concerns: spy on the Soviets without their permission, or negotiate an agreement to monitor each other's military activity. President Eisenhower tried both.

In a 1955 meeting of international leaders, Eisenhower made an "open-skies" proposal to permit Soviet and American reconnaissance flights over each other's territory. The Soviets turned him down. By the late 1950s, the limitations of reconnaissance from aircraft and balloons as well as this failure of

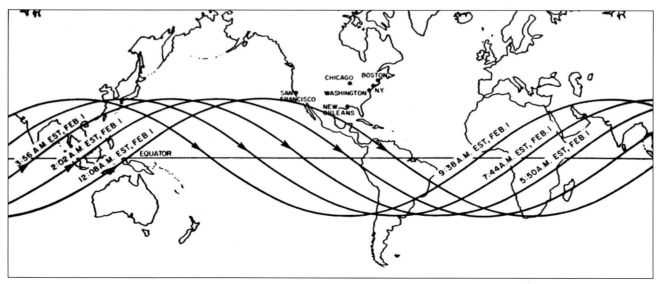

Orbit track of IGY satellite, IGY Bulletin, March 1958

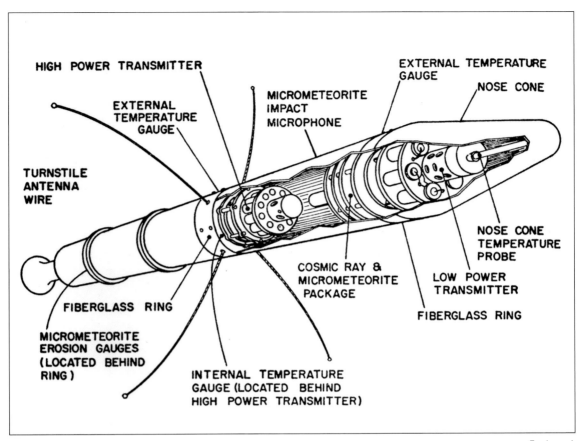

HIGH POWER TRANSMITTER

EXTERNAL TEMPERATURE GAUGE

EXTERNAL TEMPERATURE GAUGE

MICROMETEORITE IMPACT MICROPHONE

NOSE CONE

TURNSTILE ANTENNA WIRE

NOSE CONE TEMPERATURE PROBE

COSMIC RAY & MICROMETEORITE PACKAGE

LOW POWER TRANSMITTER

FIBERGLASS RING

FIBERGLASS RING

MICROMETEORITE EROSION GAUGES (LOCATED BEHIND RING)

INTERNAL TEMPERATURE GAUGE (LOCATED BEHIND HIGH POWER TRANSMITTER)

Explorer I

diplomacy created an opening for a new technology—spy satellites.

Aircraft probed Soviet territory throughout the 1950s, as (for a short time) did camera-equipped balloons. The U-2 spy plane was especially designed for reconnaissance missions. In May 1960 a U-2 was shot down over the Soviet Union. The U.S. pilot was captured, tried, and imprisoned for espionage.

Freedom of Space

The use of reconnaissance satellites raised a sensitive question under international law: was space free to all, like the open seas, or was it part of a nation's sovereign territory, like airspace?

President Eisenhower and his advisors sought to ensure international acceptance of the freedom of space. They planned to use the 1957 International Geophysical Year, a worldwide cooperative scientific study of the Earth, to set this precedent.

Eisenhower decided that a science satellite—already part of the IGY plan and less controversial than a spy satellite—would make the United States' first foray into space. Launches of the Soviet Union's Sputnik in late 1957 and the U.S. Explorer 1, a science satellite, in January 1958 were the first steps toward freedom of space.

In early 1958, a few months after the Soviets launched the first Sputnik, President Eisenhower authorized a top-priority reconnaissance satellite project jointly managed by the Central Intelligence Agency (CIA) and the U.S. Air Force. It was to launch into orbit a camera-carrying spacecraft that would take photographs of the Soviet Union and return the film to Earth.

The secret spy satellite was dubbed Corona by the CIA. To disguise its true purpose, it was given the cover name Discoverer and described as a scientific research program.

From 1960 to 1972, more than 100 Corona missions took over 800,000 photographs. As cameras and imaging techniques improved, Corona and other high-resolution reconnaissance satellites provided increasingly detailed information to U.S. intelligence analysts.

After a series of failures, the thirteenth Discoverer/Corona mission was successful. A satellite was launched and a return capsule was retrieved from orbit for the first time in August 1960. A week later, Discoverer 14 carried a camera into orbit and returned a capsule containing the first U.S. photographs of Soviet territory taken from space.

The first photograph of a Soviet military site taken from a spacecraft shows a Siberian air base at Mys Shmidta near the Chukchi Sea. It reveals objects about forty feet across, as seen from an altitude of more than a hundred miles. Film retrieved from Discoverer 14 covered more Soviet territory than all the earlier U-2 aircraft flights combined.

Soviet airfield: Discoverer 14 image, 18 August 1960

94

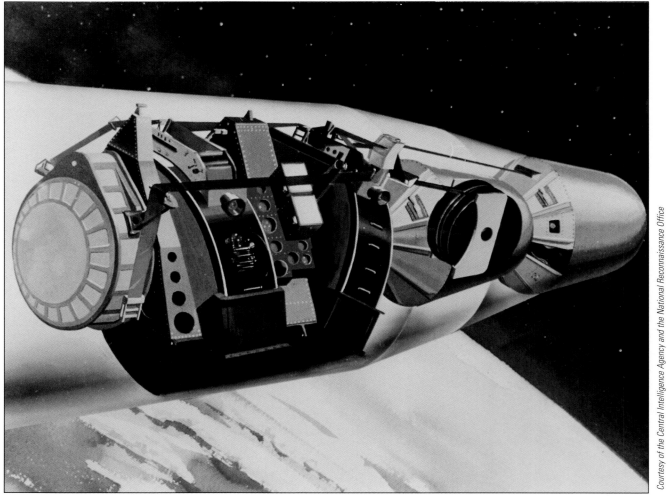

Courtesy of the Central Intelligence Agency and the National Reconnaissance Office

Artist's conception of Corona in space

Corona's Mission

The Corona project's goal was to take detailed photographs of large areas of the Earth from orbit. To do this, the project team had to meet three major technical challenges:

Design a camera able to take high-quality photographs from an altitude of more than 100 miles while moving faster than 17,000 miles per hour relative to the ground.

Point the camera and keep it steady while taking sharp pictures of specific sites.

Return the exposed film safely to Earth.

Dozens of companies and many thousands of people labored in secret to develop and operate the technologies for Corona.

This camera is a reconstructed KH-4B, the type used in the Corona project during the late 1960s and early 1970s. Objects as small as six feet across could be seen in photographs taken by this camera. It was declassified and donated to the Smithsonian Institution in 1995.

Gift of the National Reconnaissance Office

KH-4B technical specifications:
   Camera manufacturer: Itek Corporation
   Launch vehicle: Thor-Agena
   Vehicle manufacturer: Lockheed

The Corona camera used a stereo vision technique that helped CIA photographic analysts better estimate the dimensions of missiles and other objects. Two cameras were mounted side by side, one pointing slightly ahead in the satellite's direction of motion, the other slightly behind. Both took pictures of the same territory, but from different angles and with a slight delay.

This photograph, showing the Kremlin in Moscow, was taken by a Corona satellite in 1970. It is possible to distinguish cars from trucks, as well as a line of people waiting to enter Lenin's Tomb in Red Square.

Courtesy of the Central Intelligence Agency and the National Reconnaissance Office

This Corona image of the Pentagon, site of U.S. military headquarters, shows how much detail early spy satellites could reveal.

Next page: When the satellite's main camera snapped a picture of the ground, two small cameras took a picture of the Earth's horizon at the same time on the same piece of film. The horizon cameras helped interpreters calculate the position of the spacecraft relative to the Earth and verify the geographical area covered in the photo. This photo with horizon images at each end is of Luke Air Force Base in Arizona.

Horizon images and ground swath, made by Corona

Courtesy of Itek Corporation

C-119 recovering capsule

With the exposed film wound onto the reel, the nose cone return capsule was ejected at a predetermined time and location. As the capsule descended to Earth, a heat shield protected it and a radio beacon indicated its position.

At about 60,000 feet, a parachute deployed and a recovery aircraft snagged the capsule in mid-air. If the aircraft missed, the capsule was designed to land in the ocean, float briefly, and then sink to prevent recovery by the Soviets.

Later versions of tkhe Corona camera exposed twice as much film and had two return capsules for each mission.

A special Air Force unit based in Hawaii, the 6593rd Test Squadron, retrieved the Corona capsules. The squadron flew modified C-119 and C-130 aircraft, trailing a "trapeze" bar with hooks to snag the capsule's parachute.

This capsule from Corona 1117, the 122nd and final mission, returned from orbit on May 25, 1972. Corona has since been replaced by more advanced reconnaissance satellites that remain classified.

Corona's first successful mission in August 1960 marked a turning point in the Cold War. Before Corona, information about the location, number, and capabilities of Soviet weapons was limited. Although constrained by cloud cover, nighttime, and camouflage, Corona gave the United States a steady stream of photographs covering the entire Soviet Union.

These photographs allowed America's leaders to better gauge the Soviet threat. One Corona veteran said, "It was as if an enormous floodlight had been turned on in a darkened warehouse."

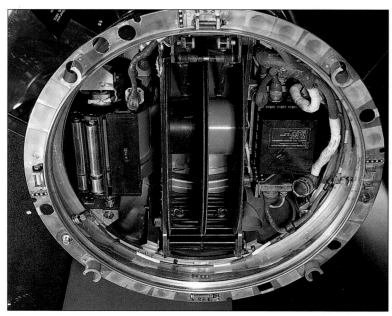

Manufacturer: General Electric. Gift of the National Reconnaissance Office

Courtesy of the Central Intelligence Agency and the National Reconnaissance Office

This 1962 photograph shows the SS-7 missile base at Yurya in Russia, the first Soviet ICBM complex to be identified in Corona images.

Debunking the Missile Gap

In the late 1950s, Soviet Premier Nikita Khrushchev had boasted that his country was turning out ICBMs "like sausages." Especially after Sputnik, many in the West believed that the Soviets might already have an arsenal of ICBMs, creating a dangerous "missile gap" between the United States and the Soviet Union.

Corona photographs proved that the "gap" actually favored the United States, which had deployed more ICBMs than the Soviet Union by the early 1960s. For America's leaders, this knowledge substantially eased fears of a Soviet surprise attack.

Photographs from the Corona missions enabled the United States to count the number of bombers and missiles in the Soviet arsenal, and also alerted American experts to nuclear weapons tests and pending space launches. Above is a 1960 Corona image of a Soviet launch complex, the Baikonur Cosmodrome.

Zenit-2

Courtesy of Soyuzkarta

The Capitol Mall, Washington, D.C. This color photograph of Washington, D.C., was taken by a Zenit satellite. Objects less than seventeen feet across can be distinguished.

Zenit: The Soviet Corona

By the time the first Corona successfully flew in August 1960, the Soviets were already designing their own spy satellite to fly over the United States. In April 1962, Zenit-2 ("Zenith"), a converted Vostok spacecraft carrying cameras instead of a cosmonaut, successfully returned film from space. As the United States did with Corona, the Soviet Union disguised the true purpose of the Zenit program. It claimed that the satellites were for scientific exploration and gave them the generic name of Kosmos.

Spies in Space

In the early 1960s, both superpowers explored the possibility of sending people to take reconnaissance pictures from space. The U.S. Air Force proposed "Blue Gemini" and a Manned Orbiting Laboratory for manned reconnaissance missions. In the Soviet Union, the Almaz ("Diamond") military space station and Merkur ("Mercury") spacecraft were designed for this purpose.

The United States abandoned its plans for such programs, but Soviet cosmonauts did some reconnaissance during the 1970s. Both sides relied mainly on

In the mid-1970s the Soviet Union orbited two Almaz reconnaissance space stations, Salyut 3 and Salyut 5. A capsule of this type returned exposed film to Earth from Salyut 5. Manufacturer: N.P.O. Machinostroenie.

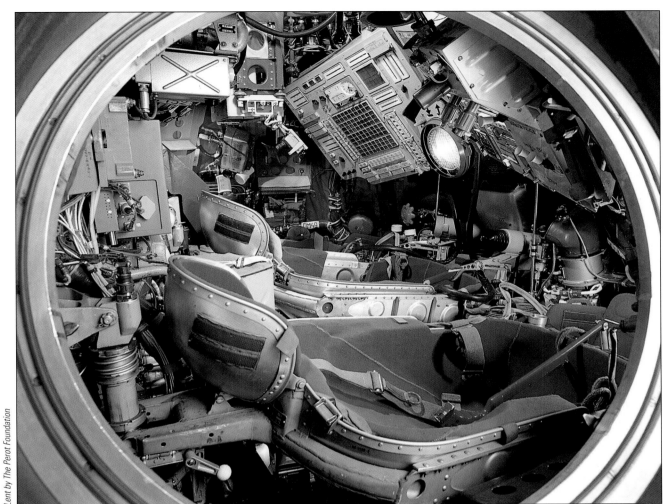

Merkur interior

less costly automated satellites to gather information about their Cold War adversaries.

This Merkur spacecraft was launched as part of an experimental military space station module, Kosmos 1443, in March 1983. The complex docked with the Salyut 7 space station, and Merkur returned five months later.

The spacecraft is fitted with seats for three cosmonauts, but it never had a crew. It was intended to ferry cosmonauts, supplies, and equipment into orbit, but the Merkur and military space station programs were terminated in favor of another program.

Merkur technical specifications:
  Length: 6 feet 7 inches
  Diameter: 10 feet
  Weight: 8,400 pounds
  Manufacturer: Central Design Bureau of Machine Building (TsKBM)
  Launch vehicle: Proton

Merkur exterior

A New Mission: Treaty Verification

The original purposes of the Corona project were to guard against a Soviet surprise attack and to provide information on targets in the Soviet Union should war occur. But as the United States and the Soviet Union entered cooperative agreements to reduce their nuclear arsenals, satellite reconnaissance unexpectedly gained a new role: watching to ensure that the treaty terms were being met.

Starting in the early 1970s, both nations began to rely on satellites to monitor compliance with arms control treaties. Reconnaissance from space now also warns of potential threats to national security throughout the world.

This image reportedly was made by a KH-11 camera. It shows a Soviet aircraft carrier under construction at a Black Sea shipyard. Under magnification, objects as small as one foot across can be seen.

Corona's Successors

In 1995, U.S. government officials ended the secrecy around Corona, opening its history to view. Details of reconnaissance satellites more recent than Corona remain officially secret, but some features are known.

Today's reconnaissance satellites are larger, comparable in size to the Hubble Space Telescope. They no longer return buckets of film from space. Instead, visual images are converted to digital data and transmitted to Earth.

Corona photoanalysts usually had to wait a week or more for photographs. Today, images from reconnaissance satellites can be received in a matter of minutes, permitting almost instant analysis of a developing threat.

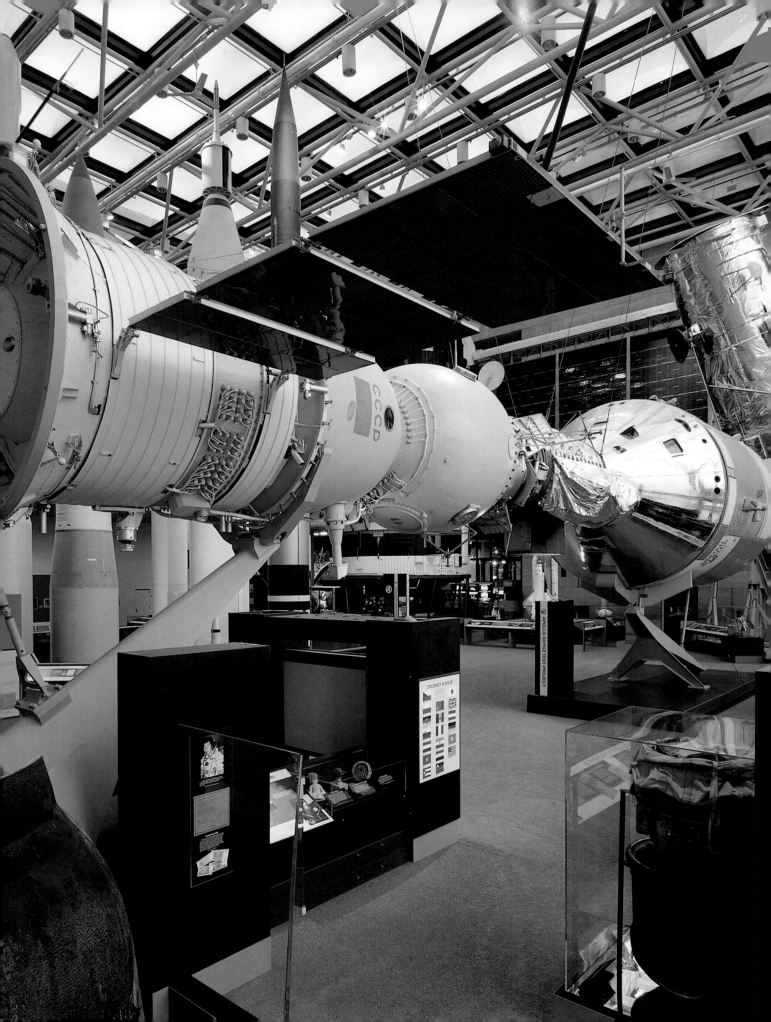

EPILOGUE

In July 1975, two manned spacecraft were launched into Earth orbit—one from Kazakhstan, the other from Florida. Their rendezvous in orbit fulfilled a 1972 agreement between the Soviet Union and the United States to participate in a joint venture in space.

The Apollo-Soyuz Test Project marked a brief thaw in the Cold War and the first time that the two rivals cooperated in a manned space mission. Engineering teams from both sides collaborated in the development of a docking module to link the spacecraft. Control centers in Moscow and Houston exercised joint duties through a cooperative exchange of tracking data and communications. The crews visited each other's spacecraft, shared meals, and worked on various tasks during several days together in space. Both sides hoped the mission might symbolize the end of competition and the beginning of an era of cooperation in space.

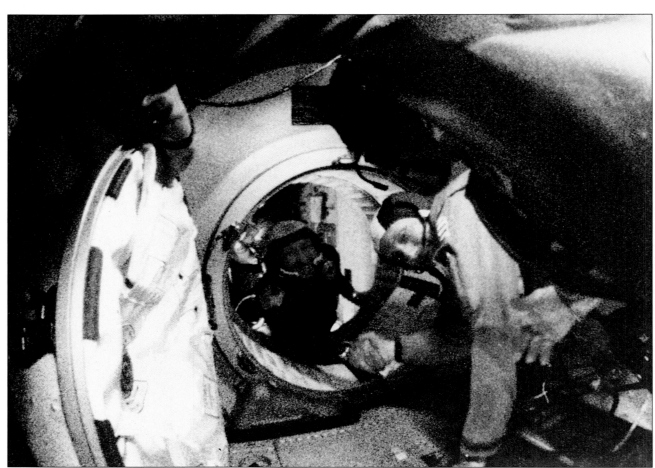

Apollo commander Thomas P. Stafford (right) and Soyuz-19 commander Aleksei A. Leonov (left) meet in space with a handshake.

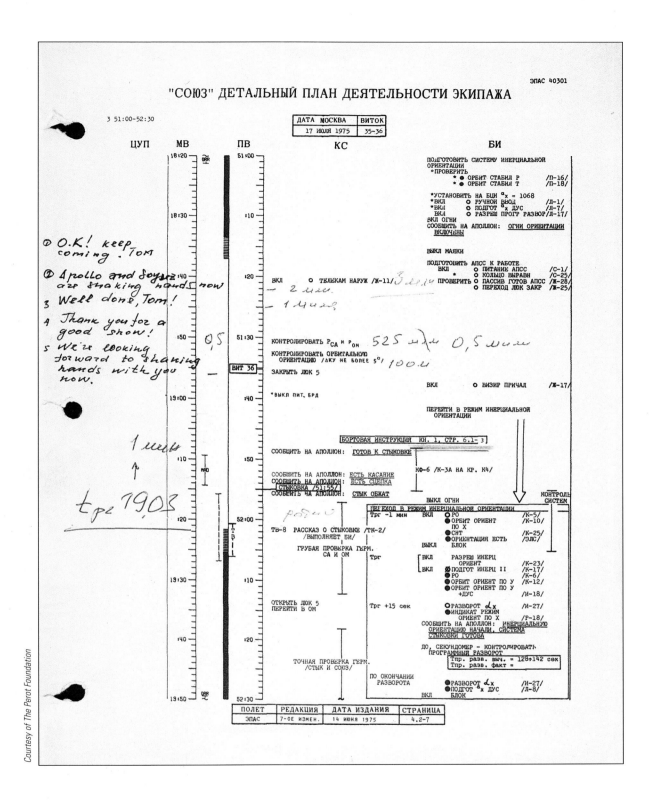

Soyuz commander Aleksei Leonov made these notes on his flight manual for the historic linkup in space. Beside the printed instructions (in Russian) for the docking maneuvers, he jotted a reminder of some English words of welcome to Apollo commander Thomas Stafford.

Artist's conception of the International Space Station

Courtesy of NASA

The dramatic launch of Sputnik in October 1957 inaugurated the Space Age. For more than a decade, through the Apollo landings on the moon from 1969 to 1972, exploration and military uses of space were shaped by the intense Cold War competition between the United States and the Soviet Union. Apollo-Soyuz suggested the possibilities of cooperation. With the end of the Cold War in the early 1990s, the United States and Russia have joined as partners in the human exploration of space through collaborative work on the Russian Mir Space Station and on the new U.S.-initiated International Space Station.

Well before Sputnik, enthusiasts saw a utopian opportunity in space exploration. Humanity's outward steps to planets and stars, some believed, would be a collective effort, an adventure of many nations and peoples, not just one or two, motivated not by fears of war but by a quest for knowledge. With the end of the Cold War, this motif has reemerged as the International Space Station starts to become a reality.

The station is the largest (and perhaps the most expensive) scientific and technological endeavor ever attempted. In addition to the United States and Russia, fourteen other nations are participating. The first segments of the station are already in orbit. Forty-five assembly flights will be required to finish the project (scheduled for 2004), requiring feats by launch vehicles and astronauts that have never before been attempted in the history of space exploration. This grand undertaking, like the race to the moon, is inseparable from U.S. foreign policy and domestic concerns. Yet as an international, cooperative, and peaceful venture, dedicated to inquiry, it stands as an emphatic conclusion to Cold War competition and as a harbinger of future exploration.